GW01157385

The Harvesting Rainwater and Water Survival Guide

Essential Prepping Strategies for Water Abundance, Self-Sufficiency, and Survival Skills in a World of Uncertainty

© **Copyright 2024 - All rights reserved.**

The content contained within this book may not be reproduced, duplicated, or transmitted without direct written permission from the author or the publisher.

Under no circumstances will any blame or legal responsibility be held against the publisher or author for any damages, reparation, or monetary loss due to the information contained within this book, either directly or indirectly.

Legal Notice:

This book is copyright-protected. It is only for personal use. You cannot amend, distribute, sell, use, quote, or paraphrase any part of the content within this book without the consent of the author or publisher.

Disclaimer Notice:

Please note the information contained within this document is for educational and entertainment purposes only. All effort has been executed to present accurate, up-to-date, reliable, and complete information. No warranties of any kind are declared or implied. Readers acknowledge that the author is not engaging in the rendering of legal, financial, medical, or professional advice. The content within this book has been derived from various sources. Please consult a licensed professional before attempting any techniques outlined in this book.

By reading this document, the reader agrees that under no circumstances is the author responsible for any losses, direct or indirect, that are incurred as a result of the use of the information contained within this document, including, but not limited to, errors, omissions, or inaccuracies.

Table of Contents

PART 1: HARVESTING RAINWATER..1
    INTRODUCTION 2
    CHAPTER 1: THE BASICS OF RAINWATER HARVESTING 4
    CHAPTER 2: THE SCIENCE BEHIND PRECIPITATION..........................14
    CHAPTER 3: PICKING YOUR SPOT..28
    CHAPTER 4: DESIGNING YOUR HARVESTING SYSTEM......................41
    CHAPTER 5: STORAGE SYSTEMS - BARRELS, GUTTERS, AND TANKS..56
    CHAPTER 6: SAFETY AND FILTRATION - ENSURING CLEAN WATER FOR EVERY USE..68
    CHAPTER 7: BEYOND THE BASICS - ADVANCED SYSTEMS AND TECHNIQUES...82
    CHAPTER 8: NATURE'S BOUNTY - USES FOR YOUR HARVEST 94
    CHAPTER 9: POTABLE RAINWATER: MAKING YOUR HARVEST DRINKABLE..106
    CHAPTER 10: A SUSTAINABLE FUTURE - TECHNIQUES IN MODERN-DAY CONSERVATION 117
    CONCLUSION: A CALL TO ACTION 124
PART 2: WATER SURVIVAL GUIDE..127
    INTRODUCTION 128
    CHAPTER 1: WATER: THE ESSENCE OF LIFE...130
    CHAPTER 2: HYDRO-GEOGRAPHY 140
    CHAPTER 3: RAINWATER AND DEW COLLECTION 150
    CHAPTER 4: RAINWATER AND DEW STORAGE 160

CHAPTER 5: THE ART OF WATER PURIFICATION 169

CHAPTER 6: SNOW AND ICE: MELTING TIPS, MYTHS AND MISCONCEPTIONS 184

CHAPTER 7: CONSERVING WATER IN SCARCE REGIONS.................. 192

CHAPTER 8: SURVIVING WHILE ON THE MOVE................................. 202

CHAPTER 9: LONG-TERM WATER STORAGE AND SAFETY................ 212

CHAPTER 10: BEYOND THE BOTTLE: THE MANY USES FOR WATER... 222

BONUS CHAPTER: CHECKLISTS 232

CONCLUSION 237

HERE'S ANOTHER BOOK BY DION ROSSER THAT YOU MIGHT LIKE 239

REFERENCES 240

Part 1: Harvesting Rainwater

A Sustainable Guide to Collecting, Storing, and Utilizing Nature's Gift for Water Conservation and Self-Sufficiency

Introduction

In a world where water is an increasingly precious commodity, imagine a solution that conserves this vital resource and transforms a mundane rain shower into a sustainable lifeline for your home and garden. Welcome to the captivating realm of "Harvesting Rainwater."

A Watery Adventure

There'll come a day when you're no longer at the mercy of your water bill, your garden thrives without guzzling gallons from the faucet, and you're not left high and dry during droughts. This book is your ticket to a watery wonderland where rain becomes a partner in your journey to sustainability.

Purpose Unveiled

The primary goal of "Harvesting Rainwater" is to empower you with the knowledge and skills to harness the incredible potential of rainwater. More than a resource, rainwater is a solution, and this book is your roadmap to unlock its full potential. It's not just about water conservation but a lifestyle shift towards self-sufficiency, eco-friendliness, and a deeper connection with the environment.

Why This Book Stands Out

What sets this guide apart from the rest in the market? Simply put, it's designed with you in mind. There's no complicated jargon or intricate diagrams that make your head spin. This book is your friendly neighbor, inviting you over for a chat about rain, barrels, and sustainable living.

- **Easy to Understand:** Forget the perplexing technicalities. This book breaks down the rainwater harvesting process into bite-sized, easily digestible pieces. You don't need an engineering degree to understand the concepts presented here.
- **Great for Beginners**: Whether you're a gardening enthusiast or someone just dipping their toes into the sustainability pool, this book is the perfect starting point. It assumes no prior knowledge, guiding you from the basics to becoming a rainwater harvesting maestro.
- **Hands-On Methods and Instructions**: This isn't a theoretical schoolbook. It's a hands-on manual. Dive into practical, step-by-step instructions that turn theory into action. By the time you put it down, you'll be ready to implement your rainwater harvesting system confidently.
- **Engaging and Accessible**: Are complicated manuals collecting dust on your shelf? This one won't join them. "Harvesting Rainwater" is a page-turner, written in a style that's engaging, humorous, and accessible.

Say goodbye to water woes and join the movement towards a more sustainable future. This book is a key to a greener, more self-sufficient tomorrow. So, dive into the pages of "Harvesting Rainwater" and let the water revolution begin. Your garden, your wallet, and the planet will thank you.

Chapter 1: The Basics of Rainwater Harvesting

Rainwater has played a vital role throughout human history. From ancient civilizations to the environmentally conscious present, rainwater harvesting has woven itself into the fabric of sustainable living. In this chapter, you'll go through the fundamental concepts of rainwater collection. You'll be tracing its roots in history, understanding its modern significance, and exploring its place within the intricate web of the natural water cycle.

Rainwater has played a vital role throughout human history.
https://www.pexels.com/photo/raindrops-1529360/

Historical Context of Rainwater Harvesting

Rain, the ageless dance of droplets from the heavens, has been an eternal companion to humanity. In the intricate choreography of nature, rainwater has been more than a fleeting visitor. It's a timeless resource that early civilizations, with their keen understanding of the environment, ingeniously harnessed for survival. Prepare to travel back in time to explore the historical context of rainwater harvesting and witness the evolution of techniques that have shaped this practice.

Unveiling the Ingenuity of the Nabataeans

The Nabataeans, inhabitants of the ancient city of Petra, were true maestros of rainwater harvesting. Situated in the heart of a desert, their survival depended on their ability to maximize every precious drop of rain. The ingenious rain gardens of Petra were a testament to their advanced understanding of water flow and conservation.

Carved into the rose-red sandstone, the Nabataean dams and cisterns formed an intricate network designed to capture and direct rainwater. These structures weren't just utilitarian. They were a marriage of functionality and artistry. The Nabataeans sculpted their environment to harmonize with the sporadic yet life-giving rains, showcasing a level of ingenuity that still captivates modern minds.

Their approach was proactive. They didn't wait for water scarcity to force innovation, anticipating the needs of their community and engineering solutions for sustained prosperity. The legacy of the Nabataeans serves as a reminder that, even in the harshest environments, humans have the potential to transform challenges into opportunities.

Greek Wisdom in Rooftop Catchment

The Greeks, renowned for their contributions to philosophy and science, also recognized the value of rainwater. In a society that esteemed wisdom, they implemented sophisticated rooftop catchment systems to capture and channel rain into storage vessels.

The Greeks understood that rain was a source of life. The rooftop catchment systems, often seen in conjunction with the architecture of ancient Greek homes, were a practical manifestation of their reverence for water. By collecting rainwater, the Greeks ensured a reliable supply for domestic and agricultural use.

This harmonious integration of practicality and philosophy reflected a holistic approach to living in concert with nature. The Greeks, while in their pursuit of knowledge, acknowledged the interconnectedness of human life with the environment. The wisdom of rooftop catchment systems wasn't just a technological achievement. It was a manifestation of a deeper understanding of the symbiotic relationship between humanity and the elements.

Mohenjo-Daro's Hydraulic Brilliance

The ancient city of Mohenjo-Daro, nestled in the fertile plains of the Indus Valley, showcases another chapter in the historical saga of rainwater harvesting. The inhabitants of this advanced civilization were pioneers in crafting intricate systems of canals and reservoirs to harness the monsoon rains.

Mohenjo-Daro's strategic water management wasn't merely about survival but about thriving. The canals and reservoirs were carefully planned and engineered systems that sustained agricultural pursuits. The brilliance of Mohenjo-Daro lay in the architectural layout of the city and the foresight to leverage the seasonal abundance of rain for long-term prosperity.

The city's advanced hydraulic systems were a testament to a civilization's organizational and engineering prowess that flourished in harmony with its environment. The practical lessons from Mohenjo-Daro transcend time, reminding you that sustainable water management is not a modern concept but an age-old wisdom rooted in shared human history.

Roman Mastery: Aqueducts and Cisterns

The Romans, synonymous with engineering marvels, elevated rainwater harvesting to an art form. The grandeur of their aqueducts and cisterns supplied water for domestic use and played a crucial role in sustaining the expansive Roman Empire.

With their impressive arches spanning across landscapes, aqueducts were engineering feats that transported water over vast distances. Cisterns, strategically placed within cities and estates, stored rainwater for times of need. The Romans recognized that rainwater was a strategic asset that could be managed on a grand scale.

The mastery of the Romans extended beyond conquests. It encompassed the thoughtful utilization of natural resources. Their aqueducts and cisterns were conduits of sustainability, ensuring a stable water supply for a burgeoning civilization. The legacy of Roman rainwater harvesting is a testament to the enduring impact of farsighted environmental stewardship.

Evolution into the Middle Ages

As the medieval period unfolded, monasteries became hubs of innovation in water management. Vast rooftop systems collected rainwater, serving agricultural and domestic needs. Monks, often the custodians of knowledge and wisdom, recognized the value of rainwater for sustenance and the spiritual and communal well-being of their societies.

In the monastic tradition, rainwater harvesting went from a practical necessity to a spiritual practice. Monasteries often featured intricate systems of gutters and downspouts that directed rainwater into storage facilities. The collected rainwater, considered pure and untainted, was used for various purposes, including brewing and medicinal preparations.

The monastic approach to rainwater harvesting reflects a profound understanding of the interconnectedness of physical and spiritual well-being. More than survival, it was about holistic living. The echoes of medieval rainwater harvesting resonate in the quiet courtyards of monasteries, where the timeless practice merged with a deeper appreciation for the sanctity of water.

Renaissance Refinement

The Renaissance period witnessed a refinement of rainwater harvesting systems. Elaborate designs adorned the estates of the wealthy, reflecting a practical approach to water conservation and an aesthetic integration of functionality and beauty. The grandeur of these systems mirrored the cultural and artistic aspirations of the time.

As the Renaissance unfolded, a renewed interest in classical knowledge and a celebration of human potential spurred advancements in various fields. In the realm of rainwater harvesting, this era saw the fusion of artistic sensibilities with practical utility. Rooftop structures became ornate, featuring intricate carvings and designs that transformed the functional elements into works of art.

The refinement of rainwater harvesting during the Renaissance was a cultural expression. The estates of the wealthy became showcases of both technological prowess and artistic ingenuity. The convergence of beauty and utility in rainwater harvesting systems mirrored the broader Renaissance spirit. It was an era when human achievement and all its facets were celebrated.

Resurgence in the Modern Era

Fast forward to the present, and you'll find yourself grappling with the challenges of a rapidly changing climate. The ancestral wisdom, however, has not been forgotten. There is a resurgence of interest in these ancient practices as modern societies seek sustainable solutions to contemporary problems. The echoes of rain gardens, rooftop catchments, and aqueducts from centuries past resonate as humans explore ways to harmonize needs with the environment.

In an era marked by technological advancements and a growing awareness of environmental concerns, the principles of rainwater harvesting are experiencing a renaissance. Water scarcity, climate change, and increasing urbanization have prompted a revisit to age-old practices that stood the test of time.

The resurgence of interest isn't merely a nostalgic look back. It's a strategic response to contemporary issues. Rainwater harvesting, once a necessity born out of survival, is now a choice. It's an informed decision to adopt sustainable water practices. The ancient techniques that allowed civilizations to flourish in diverse environments are becoming guiding lights in the quest for resilient and water-conscious communities.

Reviving Ancient Wisdom

The historical context of rainwater harvesting is an excellent example of human ingenuity and environmental stewardship. The lessons from the Nabataeans, Greeks, Indus Valley Civilization, and Romans are not relics of a bygone era. They are beacons guiding you toward a more sustainable future.

In your quest to address water scarcity and environmental challenges, you can draw inspiration from the evolution of rainwater harvesting techniques. The same principles that allowed ancient civilizations to thrive in diverse landscapes inform the contemporary efforts to build resilient and water-conscious communities.

As humanity stands at the crossroads of history and progress, the resurgence of interest in rainwater harvesting represents more than a nod to tradition. It's a conscious choice to embrace the wisdom of the past in shaping a sustainable and water-secure future. The droplets that fell on ancient civilizations continue to echo through time. They invite you to harness the liquid gold from the skies for the well-being of this planet and generations to come.

The Modern Impetus

In the unfolding narrative of the 21st century, the world finds itself on the brink of a water crisis. As populations increase, urban landscapes expand, and the capricious effects of climate change manifest with traditional water sources straining under pressure. In this era of uncertainty, rainwater harvesting emerges as a beacon of hope. It's a sustainable solution offering a reliable alternative to conventional water supplies.

Dwindling Water Resources

Water scarcity looms over the horizon in an era marked by relentless urbanization and a growing global population. Traditional water sources, like rivers and aquifers, face unprecedented stress. The water demand has surged to unparalleled levels, driven by the needs of industries, agriculture, and growing urban settlements. As these traditional sources strain to meet the demand, this water crisis necessitates innovative and sustainable alternatives.

The Significance of Rainwater Harvesting:

- **Sustainability**: Rainwater harvesting provides a sustainable alternative to overexploited traditional water reservoirs.
- **Replenishable Source:** Rainwater is a replenishable source that eases the burden of depleting water resources.
- **Conservation of Groundwater**: By capturing rainwater, you contribute to conserving precious groundwater and surface water reserves.

Environmental Responsibility

The modern impetus for rainwater harvesting extends beyond a response to water scarcity. It aligns with the growing wave of environmental responsibility sweeping across individuals and communities. As awareness of ecological issues deepens, people seek tangible ways to reduce their ecological footprint. Rainwater harvesting emerges as a tangible and impactful solution, presenting an opportunity to conserve water resources while minimizing the environmental impact associated with traditional water extraction methods.

The Environmental Impact of Rainwater Harvesting:
- **Reduced Ecological Footprint**: Rainwater harvesting reduces reliance on traditional water sources, minimizing the environmental impact of water extraction.
- **Preservation of Natural Ecosystems:** Every drop collected preserves natural ecosystems, keeping rivers undisturbed, aquifers naturally recharged, and aquatic habitats in balance.
- **Conscious Contribution**: Choosing rainwater harvesting is a conscious contribution to the broader sustainability of the planet.

Self-Sufficiency

The desire for self-sufficiency acts as a powerful motivator, prompting many to explore the realm of rainwater harvesting. By capturing rainwater on their premises, individuals gain a degree of independence from municipal water supplies. This newfound autonomy offers a reliable water source and contributes to reducing the burden on centralized water distribution systems.

Self-Sufficiency through Rainwater Harvesting:
- **Autonomy from Municipal Supplies**: Rainwater harvesting provides individuals with a reliable water source, reducing reliance on municipal supplies.
- **Agricultural Independence:** Rainwater becomes a valuable asset in agriculture, fostering self-sufficiency in nurturing crops and sustaining livestock.
- **Community Resilience:** The ethos of self-sufficiency extends to entire communities, reducing reliance on external water sources and fostering a more sustainable and resilient way of life.

A Holistic Approach to Water Security

The modern impetus for rainwater harvesting is multifaceted, addressing immediate water scarcity concerns and embracing environmental responsibility in the pursuit of self-sufficiency. As you navigate the complexities of the 21st century, rainwater harvesting emerges as a technological solution and a holistic approach to water security. This approach harmonizes with the environment, preserves natural ecosystems, and empowers individuals and communities to take charge of their water future.

The Holistic Vision of Rainwater Harvesting:
- **Harmonizing with the Environment**: Rainwater harvesting harmonizes with the environment, preserving natural ecosystems and contributing to the sustainability and preservation of the planet.
- **Empowering Individuals and Communities**: By choosing rainwater harvesting, individuals and communities progress toward a more sustainable and water-secure world.
- **A Flourishing Tomorrow:** Rainwater becomes a source of empowerment, fostering a connection between humanity and nature, shaping a flourishing and water-wise tomorrow.

Rainwater in the Natural Water Cycle

To truly appreciate the art of rainwater harvesting, you must first immerse yourself in the poetic process of the natural water cycle. This intricate choreography unfolds with the sun's tender embrace, coaxing moisture from oceans, lakes, and rivers into the sky through the enchanting process of evaporation.

The Dance of Droplets

The journey of rainwater begins high above, where the sun's warmth becomes a catalyst for transformation. Oceans, lakes, and rivers surrender their liquid essence to the sky, rising as invisible water vapor. As water vapor ascends, it merges into clouds. This cloudy collaboration is a testament to nature's artistry, a prelude to the grand performance that awaits.

The clouds gather and disperse, carrying the promise of life-giving rain. The process continues as these clouds weave intricate patterns influenced by atmospheric currents and temperature variations. When conditions align, the clouds release their gathered moisture in a cascade of rain. Isn't it a sublime spectacle that sustains life on Earth?

Cloudy Collaboration

This collaboration reaches its crescendo as clouds release their watery payload in the form of droplets. Rainfall is a fundamental act in nature's recycling system. Raindrops descend to the earth, refreshing the land, replenishing rivers, and refilling aquifers.

The rain-saturated earth becomes a stage for life's renewal. Seeds sprout, rivers flow, and ecosystems flourish in response. The rain's journey, however, is far from over. Its impact resonates in a continuous cycle, sustaining life and maintaining the delicate equilibrium of this planet.

Raindrops on leaves and soil cleanse the atmosphere and wash away dust and pollutants. The earthy fragrance that arises when rain meets dry soil is a testament to this purifying dance. The very act of rain falling to the ground is nature's way of rejuvenating and purifying the environment.

The Resilience of Rainwater

Unlike water from traditional sources, rainwater carries a simplicity that makes it an appealing alternative for various purposes. Its innate softness renders it ideal for nurturing plants, while its lack of mineral content deems it preferable for certain domestic uses.

The journey from sky to earth imbues rainwater with a unique character. As it descends, it acts as nature's purifier, cleansing itself of impurities acquired during its voyage. This innate resilience and purity make rainwater a versatile resource. It's a liquid canvas awaiting human ingenuity to paint its purpose.

The very composition of rainwater, with its lack of mineral content and low levels of dissolved solids, distinguishes it from other water sources. This purity makes it suitable for irrigation and household use. Its purity also positions it as an ideal source for certain industrial applications.

Rainwater, being free from the impurities found in ground or surface water, reduces the need for complex filtration processes. This simplicity in composition enhances its usability, all while reducing the energy and resources required to make it usable for various purposes.

The Sustainability Quotient

Rainwater harvesting isn't just about human intervention. It's a harmonious integration into this natural symphony. Collecting rainwater allows you to actively participate in the cycle without disrupting its delicate balance. It's a sustainable choice that acknowledges the interconnectedness of all elements in Earth's ecological theater.

The sustainability quotient of rainwater harvesting lies in its practical applications and alignment with the rhythms of nature. It's a choice that transcends individual needs. It reflects the ancient wisdom of civilizations that understood the droplets long before modern aspirations.

Harmony with Nature

In the grand narrative of this planet, rainwater plays a crucial role as both performer and protagonist. Its journey, from vapor to cloud to raindrop, is a testament to the resilience and interconnectedness of Earth's systems. By embracing rainwater harvesting, you embrace a harmonious relationship with nature. This relationship goes beyond mere resource utilization to a profound understanding and stewardship of the intricate water cycle.

As you collect rainwater, you become the choreographer of a sustainable future. Each collected raindrop is a step toward preserving the environment's delicate balance. With its simplicity and resilience, rainwater invites you to join the orchestra of conscious living, where every action contributes to the well-being of the planet you call home.

Integrating rainwater harvesting into your daily life is a declaration that you are not separate from nature but an integral part of its rhythms. This practice aligns with the principles of permaculture. Its philosophy mimics natural ecosystems to create sustainable and regenerative habitats suitable for humans.

The simplicity of rainwater harvesting systems, often consisting of gutters, downspouts, and storage containers, mirrors the elegance of nature's processes. This simplicity, coupled with its profound impact on local water resources, reinforces the idea that sustainability is not about complex solutions but about working with the inherent gifts of the natural world.

In the following chapters, you'll dive deeper into the practical aspects of rainwater harvesting. From the tools needed to the step-by-step processes, you will become an active participant in the age-old tradition of collecting and utilizing nature's liquid gold. As you unveil the secrets of rainwater harvesting, you'll discover the power to shape a more sustainable and self-sufficient future, one raindrop at a time.

Chapter 2: The Science Behind Precipitation

From the gentle drizzle that nurtures the soil to the torrential downpour that shapes landscapes, the science behind precipitation is a fascinating journey through the heart of the water cycle. This chapter unravels the foundational meteorological principles governing rain formation and fall, exploring the intricate dance of water molecules as they traverse the atmosphere.

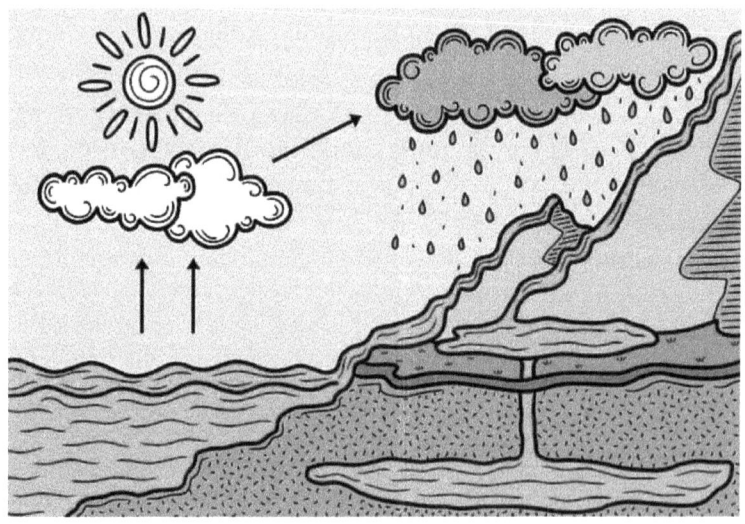

The water cycle, a mesmerizing process orchestrated by the forces of nature, is a perpetual dance that sustains life on this planet.

https://pixabay.com/zh/illustrations/water-cycle-rain-clouds-8176128/

The Water Cycle Demystified

The water cycle, a mesmerizing process orchestrated by the forces of nature, is a perpetual dance that sustains life on this planet. At its core are three captivating acts: evaporation, condensation, and precipitation. Here's a closer look at the intricacies of each stage to demystify the awe-inspiring journey of water molecules as they traverse the vast expanse of the atmosphere.

Evaporation

At the heart of the water cycle lies the enchanting process of evaporation. This act unfolds under the tender caress of the sun's warming rays. Water, in its liquid form, experiences a magical transformation into vapor. This metamorphosis is more than a scientific phenomenon. It is a fascinating dance of molecules. It's a poetic interplay between the liquid surface of oceans, lakes, and rivers and the beckoning call of the sun.

The Dance of Molecules

- **Solar Embrace:** The sun extends its golden fingers across the Earth's surface, imparting kinetic energy to water molecules. This solar embrace is where water molecules gain the energy to liberate themselves from liquid form.
- **Escape to the Skies:** With newfound energy, water molecules shed their liquid form and ascend into the atmosphere. This ethereal ascent marks the start of a journey that transcends geographical boundaries and embraces the boundless expanse above.
- **Global Voyage:** Once liberated, these water vapor molecules go on a global voyage, carried by air currents and wind. From the balmy tropics to the frigid poles, the liberated vapor becomes an intrepid traveler, ready to engage in the next act of the water cycle.

The Atmospheric Odyssey

- **Air Currents and Wind:** The liberated water vapor becomes a passenger on air currents and wind, creating a dynamic aerial phenomenon. These atmospheric currents carry water vapor across vast distances, shaping the atmospheric dynamics that influence weather patterns.

- **Moisture Reservoirs:** The vapor, now suspended in the atmosphere, forms moisture reservoirs that hold the potential for future precipitation. These reservoirs, invisible to the naked eye, are essential contributors to the delicate balance that sustains life on Earth.
- **Interconnected Systems:** The atmospheric odyssey of water vapor is part of a complex and interconnected system that influences climate, weather, and the distribution of water resources across the globe.

Condensation

As the vapor molecules ascend into the atmosphere, they encounter cooler air at higher altitudes. This encounter triggers a majestic transformation, the act of condensation. In it, vapor surrenders its ephemeral form, condensing into tiny droplets or ice crystals. The newly formed water particles then gather around particles like dust or aerosols, combining to create the canvas upon which clouds paint their ethereal beauty across the sky.

The Symphony of Condensation

- **Temperature:** The change in temperature at higher altitudes is the catalyst of this stage. Cooler air encourages the vapor molecules to slow down and embrace their liquid form once again.
- **Gathering in Clouds**: As condensation takes hold, these minuscule water particles dance around atmospheric particles, forming clouds. These clouds, in their myriad shapes and sizes, become the visual poetry of the sky, reflecting the essence of the atmosphere.
- **Aerial Artistry:** The resulting clouds, whether wispy cirrus or dense cumulus, capture and reflect the ever-changing moods of the atmosphere. This act of condensation decorates the sky and sets the stage for the grand finale of precipitation.

Cloud Formations and Aesthetic Revelry

- **Diverse Cloud Types**: Condensation gives rise to an array of cloud types, each with its unique characteristics. Cirrus clouds are high and wispy, while cumulonimbus clouds are towering and majestic, heralding the potential for intense precipitation.

- **Weather Indicators**: Cloud formations serve as invaluable indicators of imminent weather changes. Understanding the nuances of cloud aesthetics allows meteorologists and weather enthusiasts to decipher atmospheric conditions and predict upcoming precipitation events.
- **Artistic Splendor:** The aesthetic revelry of cloud formations is a testament to the creative artistry of nature. From radiant sunsets reflected in altocumulus clouds to the ominous beauty of an approaching storm in nimbostratus clouds, condensation transforms the sky into a canvas of ever-changing masterpieces.

Precipitation

The grand culmination of the water cycle's atmospheric journey is the act of precipitation. It occurs when the condensed water droplets within clouds grow heavy enough to overcome the resistance of air currents. Under the influence of gravity, they descend earthward, transforming into varied forms of precipitation, including rain, snow, sleet, or hail.

The Dramatic Descent

- **Growing Heavier:** Within the clouds, the water droplets continue to grow in size as they collide and merge. This growth transforms them into precipitation – ready to make its descent.
- **Gravity's Pull:** The moment arrives when these condensed droplets become too weighty for the air to support. Gravity, the omnipotent force, pulls them downward, initiating the descent that defines precipitation.
- **Elixir of Life:** As these droplets kiss the Earth, they contribute to the vital cycle of life. Whether nurturing the soil, replenishing lakes and rivers, or sustaining ecosystems, precipitation is the elixir that rejuvenates and sustains this planet.

The Holistic Impact

- **Soil Nourishment:** Precipitation seeps into the soil, providing essential hydration to plant roots. This nourishment is fundamental to the growth and vitality of terrestrial ecosystems.
- **Aquatic Replenishment:** Lakes, rivers, and oceans receive a replenishing embrace from precipitation. This influx of freshwater sustains aquatic habitats, maintaining the delicate balance of marine ecosystems.

- **Ecosystem Resilience:** The holistic impact of precipitation extends beyond individual components of the environment. It contributes to the resilience of ecosystems, ensuring the continued vitality and diversity of life on Earth.

The Harmonious Cycle

The journey from evaporation to precipitation is not just a linear progression. It's a harmonious cycle that perpetuates life on Earth. Each act in this atmospheric dance is interconnected, creating a seamless choreography that repeats itself endlessly. From the liquid embrace of evaporation to the artistic formations of condensation and the dramatic descent of precipitation, the water cycle is a living, breathing testament to the interconnectedness of nature.

The Enchanting Finale

As you demystify the water cycle, you unveil the intricate beauty that sustains life on this planet. This perpetual dance, conducted by the sun, the atmosphere, and the Earth, is a testament to the resilience and interconnectedness of nature. As you gaze upon the clouds, feel the rain on your face, and witness the cycle unfold around you, you are not a mere spectator but an active participant in the grand symphony that is the water cycle.

Factors Influencing Rainfall

In the intricate workings of rainfall, nature conducts a symphony where various factors harmonize to create the delicate dance of precipitation. Each element plays a crucial role, from temperature fluctuations and the topographical stage upon which rain unfolds to the nuanced air currents and humidity that set the rhythm. In this section, you'll learn about the complex interplay of factors influencing rainfall dynamics, deciphering the poetry written in raindrops.

Temperature Fluctuations

Temperature holds the baton that directs the rhythm of precipitation patterns. Its influence is profound, shaping the atmospheric conditions that give rise to rainfall. The balance between temperature fluctuations is a key to deciphering the dynamics of rain.

The Dance of Warm and Cool

- **Warmer Air, Increased Moisture:** In the atmosphere, warmer air leads to increased moisture retention. As temperatures rise, air gains the capacity to hold more water vapor through evaporation. It sets the stage for heightened evaporation from oceans, lakes, and other water bodies, fostering the birth of clouds.

- **Cooler Temperatures, Condensation Groundwork:** On the contrary, cooler temperatures provide the backdrop for condensation to take center stage. When warm, moisture-laden air encounters cooler conditions, it reaches its dew point (the temperature at which condensation occurs). This transition from vapor to liquid lays the groundwork for cloud formation and, eventually, precipitation.

Understanding the balance between these temperature fluctuations unveils the intricate dance of rainfall dynamics. From the initial evaporation to the eventual condensation and precipitation, temperature is the guiding force that shapes the symphony of rain.

Topography

The Earth's topography is the grand stage upon which precipitation patterns unfold, adorned with geographic nuances that add depth to the rainfall symphony. Mountains, valleys, and plains interact with air masses, influencing their ascent or descent and shaping the spatial distribution of rainfall.

Mountainous Terrain

- **Ascent and Enhanced Condensation:** Mountains play a pivotal role in the rainfall narrative. As moist air ascends a mountain range, it undergoes adiabatic cooling. The cooling process enhances condensation, transforming the ascending air mass into clouds. This phenomenon results in increased rainfall on the windward side of the mountain.

- **Leeward Side and the Rain Shadow Effect:** On the leeward side of the mountain, a contrasting scenario unfolds. As air descends, it undergoes adiabatic warming, creating conditions less favorable for condensation. This leeward side experiences a rain shadow effect, characterized by drier conditions and reduced rainfall.

Valleys and Plains

- **Influence on Air Mass Movements:** Valleys and plains, while not as topographically imposing as mountains, also influence rainfall patterns. They guide the movement of air masses, facilitating the ascent or descent that contributes to the spatial distribution of

precipitation.
- **Interplay with Atmospheric Dynamics**: The interplay between topography and atmospheric dynamics creates a multifaceted stage for rainfall. The topographical features become integral components of the atmosphere, influencing the intensity and distribution of precipitation.

Air Currents

The movement of air currents across the globe shapes precipitation patterns with finesse. Trade winds, prevailing westerlies, and polar easterlies dictate the movement of air masses, influencing where precipitation occurs. Convergence zones, where air masses collide, become focal points for intense rainfall.

Trade Winds

- **Equatorial Convergence Zone**: Trade winds converge near the equator, creating the equatorial convergence zone. That becomes a breeding ground for intense rainfall. The warm, moist air rises, cools, and condenses, giving birth to the lush rainforests that characterize equatorial regions.
- **Tropical Rain Belts:** Trade winds, in their easterly course, also contribute to the formation of tropical rain belts. These bands of concentrated rainfall encircle the Earth, creating the climatic conditions that support diverse ecosystems.

Prevailing Westerlies and Polar Easterlies

- **Mid-Latitude Dynamics**: Prevailing westerlies dominate the mid-latitudes, and polar easterlies influence high latitudes, contributing to the mid-latitude dynamics of rainfall. These air currents guide weather systems, influencing precipitation patterns in temperate regions.
- **Storm Tracks and Frontal Boundaries**: The convergence of air masses along frontal boundaries, influenced by prevailing westerlies, becomes a theater for dynamic weather patterns. Storm tracks, shaped by these air currents, become corridors of intense rainfall.

Understanding these atmospheric currents unveils the intricate choreography of rain distribution on a planetary scale. The movements of air masses, driven by the Earth's rotation and solar heating, create a dynamic interplay that orchestrates rainfall in diverse regions across the globe.

Humidity

Humidity, a measure of the moisture content in the air, is a critical player in the precipitation narrative. It sets the rhythm for the dance of moisture, contributing to both the birth of clouds through evaporation and the eventual precipitation through condensation.

High Humidity Fosters Evaporation

- **Moisture-Laden Atmosphere:** High humidity levels create a moisture-laden atmosphere conducive to evaporation. Water bodies, soil, and vegetation release moisture into the air, saturating it with water vapor.
- **Evaporation from Oceans:** In regions with high humidity, like coastal areas and tropical climates, oceans play a significant role. The warm ocean surfaces provide ample moisture for evaporation, becoming the primary source for the moisture-rich air masses that fuel precipitation.

Condensation and Precipitation

- **Saturation and Condensation:** As air saturated with moisture ascends or encounters cooler conditions, it reaches its saturation point. That triggers the process of condensation, where water vapor transforms into tiny droplets or ice crystals, forming clouds.
- **Birth of Raindrops:** The condensed droplets, growing in size, become raindrops. The delicate balance between humidity and temperature determines when condensation prevails, leading to the birth of raindrops that will descend as precipitation.

Intensity and Duration of Rainfall

- **Humidity and Rainfall Intensity:** The intensity of rainfall is closely linked to humidity levels. High humidity contributes to more significant evaporation, creating the conditions for intense and prolonged rainfall events.
- **Seasonal Variations:** Humidity levels also exhibit seasonal variations, influencing the character of rainfall in different periods. Understanding these variations is crucial for deciphering the nuances of precipitation dynamics.

Deciphering the Symphony

In the grand symphony of rainfall, temperature fluctuations, topography, air currents, and humidity intertwine, creating a harmonious dance that sustains life on Earth. The interconnectedness of these factors forms a complex web, and deciphering their symphony provides insights into the diverse rainfall patterns witnessed across the globe.

The Interplay of Those Factors

- **Dynamic Relationships**: The relationship between temperature and humidity, the influence of topography on air masses, and the choreography of air currents contribute to the dynamic interplay that shapes rainfall patterns.
- **Regional Nuances**: Different regions experience unique combinations of these factors, giving rise to diverse climates and ecosystems. Each region tells a distinct rainfall story, from the monsoons in Southeast Asia influenced by oceanic and continental air masses to the temperate rainfall patterns shaped by prevailing westerlies.
- **Impact on Ecosystems:** The influence of these factors extends beyond meteorological dynamics to ecosystem health. Rainfall patterns dictate the availability of water resources, influencing the flora and fauna that thrive in specific regions.

As you delve into the factors influencing rainfall, you witness the intricate choreography of nature's ballet. From the nuanced guidance of temperature fluctuations to the dramatic topographical stage, the orchestrated movements of air currents, and the rhythmic interplay of humidity, each factor contributes to the symphony of rainfall.

Understanding this symphony is not merely an academic pursuit. It's a journey into the heart of Earth's vitality, where raindrops become the verses that narrate the story of life itself. In the ongoing narrative of the planet's water cycle, these factors continue to dance, creating the ever-changing melody of rainfall that sustains the beauty and diversity of this world.

Predicting Precipitation

In the ever-changing tapestry of this planet's climate, the ability to predict precipitation is paramount. It guides your preparedness for weather events, dictates agricultural practices, and helps people to understand Earth's water cycle. In this section, you'll journey through modern science and traditional wisdom, learning about the methods used to predict precipitation and bridge the gap between cutting-edge technology and ancestral insights.

Meteorological Forecasts

In modern science, meteorological forecasts are the guiding compass in anticipating precipitation patterns. Utilizing state-of-the-art technology, weather scientists harness the power of advanced tools to analyze vast datasets, interpret satellite imagery, and run sophisticated computer models. These tools enable them to predict atmospheric conditions, offering valuable insights into when and where precipitation will occur.

Advanced Technology at Play

- **Data Analysis:** Meteorologists delve into an extensive array of data, ranging from temperature and humidity levels to air pressure and wind patterns. Analyzing this data allows them to discern the complex interplay of factors that contribute to precipitation.
- **Satellite Imagery:** High-resolution satellite imagery provides a bird's-eye view of atmospheric conditions. It allows scientists to track cloud formations, identify weather systems, and monitor the development of potential precipitation events.
- **Computer Models:** Based on the collected data, advanced computer models simulate the atmosphere's behavior. These models consider various variables, enabling meteorologists to predict the timing, intensity, and duration of precipitation events.

Short-Term Forecasts to Extended Projections

- **Hourly and Daily Predictions:** Short-term forecasts, ranging from hourly to daily predictions, offer insights into imminent weather changes. These forecasts are crucial for planning daily activities, travel, and local events.
- **Extended Projections:** Meteorologists also provide extended projections that cover longer time frames, such as weekly or monthly forecasts. While these projections are uncertain, they offer valuable insights for mid-range planning and preparation.

Traditional Wisdom

Beyond cutting-edge technology, traditional wisdom cultivated over generations offers a unique perspective on predicting precipitation. Indigenous communities, deeply connected to the natural world, have developed a keen understanding of impending weather changes by observing natural indicators. This harmonious integration of ancestral knowledge with contemporary forecasting methods enriches the ability to foresee precipitation events.

Nature's Indicators

- **Animal Behavior:** Observing the behavior of animals has long been recognized as a reliable indicator of impending weather changes. Birds flying lower, cows lying down, or ants building their nests higher signal changes in atmospheric conditions.
- **Cloud Formations**: The art of reading cloud formations is a skill passed down through generations. Cloud types, colors, and patterns provide clues about upcoming weather. For example, the towering cumulonimbus clouds often herald thunderstorms.
- **Atmospheric Phenomena**: Natural occurrences like the halo around the moon or the red hues during sunrise and sunset have been observed for centuries as signs of changing weather. These atmospheric phenomena are woven into the fabric of traditional forecasting.

Ancestral Insights:

- **Cultural Knowledge:** Indigenous cultures often have specific cultural knowledge and rituals tied to weather predictions. This knowledge is shared within communities and plays a vital role in agricultural practices, hunting, and other aspects of daily life.
- **Interconnectedness with Nature:** Traditional forecasting emphasizes the interconnectedness between humans and nature. It recognizes that surroundings offer subtle cues about the changing rhythms of the natural world.

Rainfall Patterns and Climate Zones

Understanding the broader context of rainfall patterns in different climate zones contributes to more accurate predictions. Different regions exhibit distinct precipitation characteristics influenced by their proximity to the equator, local geography, and atmospheric dynamics. Recognizing these climatic nuances enhances the ability to predict when specific regions are more likely to experience rainfall.

Tropical Rainforests

Tropical regions near the equator experience consistent and heavy rainfall throughout the year.
https://www.pexels.com/photo/photo-of-foggy-forest-4633377/

- **Proximity to the Equator**: Tropical regions near the equator experience consistent and heavy rainfall throughout the year. The sun's direct rays at the equator create warm air, leading to the ascent of moist air masses and frequent precipitation.
- **Diverse Ecosystems:** The lush tropical rainforests are a testament to the abundance of rainfall. The consistent water supply supports diverse ecosystems, making accurate predictions crucial for managing these rich and fragile environments.

Arid and Semi-Arid Regions

- **Sporadic but Intense Precipitation:** Arid and semi-arid regions, like deserts, may experience sporadic but intense precipitation events. Understanding the factors contributing to these infrequent but impactful rainfall events is essential for water resource management.

- **Flash Flooding Risks:** In arid regions, the soil may have low permeability, leading to rapid runoff during intense rainfall. It poses the risk of flash flooding, making accurate predictions vital for mitigating potential hazards.

Temperate Climates
- **Seasonal Variations:** Temperate climates often exhibit distinct seasons with variations in precipitation. Understanding the seasonal patterns allows for better predictions regarding when rain is more likely to occur and its potential impact on agriculture and ecosystems.
- **Influence of Prevailing Winds:** Prevailing westerlies in temperate regions play a role in shaping rainfall patterns. Understanding the influence of these wind patterns contributes to accurate predictions.

Bridging the Gap

The synergy between modern science and traditional wisdom offers a holistic approach to predicting precipitation. While meteorological forecasts provide precise and data-driven predictions, traditional knowledge systems offer a nuanced understanding of nature's subtle cues. Integrating these knowledge systems enhances the ability to anticipate and adapt to changing weather conditions.

Cross-Cultural Collaboration
- **Knowledge Exchange:** Facilitating a cross-cultural exchange of meteorological knowledge enriches the collective understanding of weather patterns. Meteorologists can benefit from insights gained through traditional wisdom and vice versa.
- **Community Engagement:** Involving local communities in weather monitoring and prediction fosters a sense of ownership and empowerment. With their deep connection to the land, Indigenous communities contribute valuable observations that can complement scientific data.

Climate Resilience
- **Adaptive Strategies:** Incorporating traditional wisdom into climate resilience strategies enhances the adaptability of communities. Traditional forecasting methods, rooted in centuries of observation, offer early warnings and guide adaptive

practices.
- **Preserving Biodiversity**: Accurate predictions are crucial for preserving biodiversity in various ecosystems. Indigenous knowledge, intimately tied to the rhythms of nature, contributes to sustainable practices that protect diverse flora and fauna.

Modern science and traditional wisdom play indispensable roles in the intricate dance of predicting precipitation. Meteorological forecasts, with their cutting-edge technology and data-driven precision, provide you with valuable insights into the complex dynamics of the atmosphere. Simultaneously, traditional knowledge systems, cultivated over generations, offer a profound connection to the natural world and its subtle indicators.

In this chapter, you've navigated the intricate realms of the water cycle, unveiling the science behind precipitation. From the ephemeral journey of water molecules through evaporation to the intricate factors influencing rainfall, the dance of precipitation is a symphony conducted by nature itself. As you explore the mechanisms shaping rainfall patterns, the stage is set for a deeper understanding of this planet's atmospheric dynamics.

Chapter 3: Picking Your Spot

In the pursuit of sustainable water practices, the art of harvesting rainwater stands as a pivotal solution. Yet, its success hinges on selecting the optimal spot. From the size and material of your roof to the lay of the land and the nuances of local climate, each factor plays a role in determining water catchment efficiency. This chapter considers picking the perfect spot, providing you with insights to balance functionality, aesthetics, and environmental impact.

Rooftop Considerations

In rainwater harvesting, the rooftop takes center stage. It's where the transformation of precipitation into a valuable resource begins. The dimensions, angles, and materials of your roof play a key role in determining the volume and quality of the rainwater you can capture.

In rainwater harvesting, the rooftop takes center stage.
https://www.pexels.com/photo/photo-of-roof-while-raining-2663254/

Roof Area and Collection Efficiency

The size of your roof directly dictates the potential volume of rainwater you will collect. This catchment area, a critical metric in rainwater harvesting, is determined by accurately measuring the dimensions of your roof. Precision ensures you harness the full potential of available rainfall.

Efficiency Considerations

While larger roofs offer more substantial catchment areas, efficient use of space and consideration for the intended use of harvested water is paramount.

- **Bigger Isn't Always Better:** Larger roofs indeed provide more significant catchment areas, enhancing the potential for water collection. However, it's crucial to strike a balance. Consider the available space, your water needs, and the intended use of the harvested water.

- **Tailoring to Needs:** Assess your water requirements and storage capacity. This understanding helps you optimize your catchment area to meet your specific needs without unnecessary excess.

Additional Considerations

Expanding on the calculation of catchment area, it's essential to consider factors that might affect efficiency:

- **Roof Slope Variation**: In cases where the roof has varying slopes, calculate the catchment area for each segment separately. This nuanced approach ensures accurate estimations.
- **Obstructions and Adjustments**: Account for any obstructions on the roof, such as chimneys or skylights, which may affect water flow. Adjustments to gutters and downspouts will optimize collection efficiency.

Roof Angles and Pitch

The angles and pitch of your roof add complexity to rainwater harvesting. They influence the speed of water runoff and the efficiency of collection systems.

Optimal Pitch

The pitch of your roof, its incline or slope, is a critical factor in rainwater harvesting.

- **Moderation Is Key:** A moderate pitch is often considered optimal for rainwater harvesting. Steep pitches lead to faster runoff, reducing the time water spends on the roof. Striking a balance is crucial for maximizing collection efficiency.
- **Preventing Runoff Issues:** A moderate pitch allows water to linger on the roof for a sufficient duration, promoting effective collection. It prevents issues associated with rapid runoff, ensuring a steady flow into your harvesting system.

Adjusting for Angles

Different roof designs and angles require tailored approaches to enhance water flow and collection efficiency.

- **Flat Roofs:** Flat roofs offer larger catchment areas but may require specialized systems to optimize water flow. Design adjustments and strategic placement of gutters will compensate for variations in roof angles.
- **Gabled Roofs:** Gabled roofs, with their slopes on either side, offer effective water runoff. Making sure that gutters are well-positioned to capture water along the slopes enhances efficiency.

Additional Considerations

- **Snow Load and Pitch:** In regions experiencing snowfall, consider the pitch's impact on snow accumulation. A steeper pitch will shed snow more effectively, preventing excessive amounts.

- **Roof Material Influence**: Certain roofing materials perform optimally at specific pitches. Investigate manufacturer recommendations to align your roof pitch with the chosen materials.

Roof Materials and Water Purity

The material composing your roof isn't merely an aesthetic choice. It's a key player in the quality of harvested rainwater. Different materials either introduce contaminants or contribute to cleaner water.

Metal Roofs

Corrosion-resistant metals like zinc or aluminum are popular choices for rainwater harvesting. They minimize leaching and contribute to cleaner water.

- **Durability and Purity:** Metal roofs are durable and corrosion-resistant, ensuring longevity. They also contribute to cleaner water by minimizing the introduction of contaminants.
- **Applicability to Harvesting Systems**: Metal roofs are compatible with various rainwater harvesting systems, offering versatility in design and implementation.

Asphalt Shingles

While common in roofing, asphalt shingles introduce small particles and contaminants into harvested water. Mitigating these concerns requires strategic solutions.

- **Particle Concerns**: Asphalt shingles shed small particles, affecting the purity of harvested water. Installing a first-flush diverter helps divert initial runoff, reducing particle content.
- **Regular Maintenance:** Periodic inspection and maintenance of asphalt roofs are crucial. Cleaning gutters and roof surfaces minimize the accumulation of debris and contaminants.

Treated Wood or Composite Shingles

These materials introduce chemicals into the harvested water, necessitating careful analysis and additional filtration measures.

- **Chemical Concerns**: Treated wood or composite shingles release chemicals into harvested water. Conducting thorough research on the specific materials used leads to awareness of potential contaminants.

- **Filtration Solutions**: Implementing additional filtration systems, such as sediment filters or activated carbon filters, will further purify water harvested from roofs with treated wood or composite shingles.

Additional Considerations:
- **Regular Roof Inspections**: Periodic inspections of your roof's condition are crucial. Detecting and addressing issues such as rust on metal roofs or deteriorating shingles on asphalt roofs ensures the longevity of the roof and the quality of the harvested water.
- **Material Longevity:** Consider the lifespan of roofing materials concerning your long-term rainwater harvesting goals. Investing in durable materials aligns with sustainability and reduces the frequency of replacements.

As you journey through rainwater harvesting, consider your rooftop as the conductor of a grand performance. The dimensions, angles, and materials harmonize to create a melody of efficient water collection. Precision in measurement, thoughtful consideration of pitch, and mindful selection of roofing materials contribute to the purity and abundance of the harvested rainwater. As you fine-tune each aspect, you optimize your rainwater harvesting system and contribute to the broader movement toward water conservation and a greener, more sustainable future.

Terrain and Topography

Beneath your feet lies a tapestry of contours and slopes. The terrain and topography of your property will either facilitate the smooth flow of water toward storage or present challenges that demand strategic solutions. It's time to explore how the natural features beneath your feet shape the intricate dance of rainwater harvesting.

Slope and Water Flow

The slope of your property is a dynamic force that dictates the natural flow of water. Efficiently harnessing this flow ensures rainwater travels from catchment surfaces to storage with minimal resistance.

- **Leveraging Natural Slopes**: Consider yourself fortunate if your property boasts natural slopes. Leverage these contours to guide water towards designated collection points. It will significantly reduce the need for complex drainage systems, allowing you to embrace the simplicity inherent in nature.

- **Creating Artificial Slopes**: In scenarios where natural slopes are insufficient or non-existent, consider introducing artificial slopes through strategic landscaping adjustments. By sculpting the terrain, you can redirect water flow, enhancing overall collection efficiency.

Water Flow Paths and Collection Points

Understanding how water moves across your property requires you to decipher nature's blueprint. Identifying optimal collection points involves a thoughtful analysis of the paths water takes during rainfall events.

- **Gutter Systems as Navigators**: Well-designed gutter systems act as the navigators of this natural journey. They guide water along predetermined paths, preventing chaotic runoff. Regular maintenance is the key to ensuring gutters remain clear, preventing blockages that could impede the smooth flow of water.
- **Strategic Placement of Storage Systems:** Positioning water storage tanks or reservoirs at points where runoff naturally converges is a masterstroke. It reduces the need for extensive piping systems, tapping into the simplicity of aligning with natural water courses.
- **Utilizing Swales and Berms**: Swales are depressions designed to redirect water flow, and berms can be strategically incorporated. These natural features assist in directing water toward desired collection points, enhancing the efficiency of rainwater harvesting.

Vegetation and Structures

The presence of vegetation on your property introduces both challenges and benefits to the rainwater harvesting narrative. Trees and plants act as natural filters, reducing contaminants in harvested water. However, they also contribute to debris, necessitating additional filtration measures.

- **Balancing Act of Nature**: Embrace the dual role of vegetation. While trees and plants contribute to water purity by acting as natural filters, they shed leaves and debris, potentially affecting the cleanliness of collected water. Striking a balance involves regular maintenance and additional filtration measures.
- **Strategic Planting for Water Retention:** Thoughtful vegetation placement will enhance water retention in the soil. It aids the prevention of soil erosion and promotes a more sustained release

of water into the harvesting system.

Structures as Collection Points

Man-made structures, from sheds to outbuildings, significantly influence water flow patterns on your property. Thoughtful consideration of these elements will streamline the rainwater harvesting process.

- **Structures' Roofs as Catchment Areas**: Consider integrating existing structures into your rainwater harvesting system. Roofs of sheds or outbuildings serve as supplementary catchment areas, expanding the overall capacity for water collection.
- **Harmony of Aesthetics and Functionality**: Striking a balance between the aesthetic appeal of landscaping and the functional requirements of rainwater harvesting is an art. Thoughtful design ensures harmony between the natural environment and the infrastructure designed to capture and store rain's precious gift.
- **Utilizing Pervious Pavements**: Consider using permeable pavements in areas where hardscaping is unavoidable. These surfaces allow water to penetrate, reducing runoff and facilitating its absorption into the ground, contributing to the overall health of your rainwater harvesting system.

Navigating the Rainwater Harvesting Landscape

As you navigate the landscape of rainwater harvesting, the strategic placement of storage systems and the integration of both natural and built elements create a harmonious composition. Once perceived as static, the lay of the land beneath your feet now becomes a dynamic partner in the dance of water from the sky to storage.

Embrace the contours and slopes, work with the natural flow, and let your rainwater harvesting system become an extension of the landscape, seamlessly blending into the intricate design of your property. Nature and design, when choreographed with precision, transform rainwater harvesting from a practical necessity to a poetic interaction with the land itself. As you navigate this, remember that each slope, tree, and structure you've built contributes to a system that celebrates both the practicality and beauty of sustainable living.

Harmonizing Nature and Infrastructure

In exploring terrain and topography in rainwater harvesting, it's crucial to emphasize the profound connection between nature's dynamics and the man-made infrastructure designed to harness it. The process of rainwater harvesting unfolds most beautifully when these elements exist in harmony.

- **Adaptation to Local Conditions**: Recognize that each property is unique, and the strategies employed should be adapted to local topography, climate, and vegetation.
- **Continuous Observation and Adjustment:** As seasons change and landscapes evolve, ongoing observation and occasional adjustments to your rainwater harvesting setup ensure its continued effectiveness.
- **Educational Outreach:** Share your experiences and knowledge about terrain and topography in rainwater harvesting with your community. Encouraging sustainable practices contributes to a broader movement toward water conservation.

In this intricate dance of nature and design, your property becomes a canvas where rainwater transforms from a fleeting visitor into a cherished resident. Remember that the art of rainwater harvesting is not just about capturing water. It's about creating a sustainable and harmonious coexistence between human habitation and the natural world.

Regional and Climatic Factors

In rainwater harvesting, the final act transcends the microcosm of individual properties and delves into the macrocosm. Pay close attention to how local weather patterns and geographical features impact your harvesting decisions and potential yield. Understanding the broader climatic context will empower you to tailor your systems to regional intricacies, creating a symphony that harmonizes with nature's rhythms.

Proximity to Bodies of Water

The proximity of your location to bodies of water introduces a climatic nuance that profoundly influences rainwater availability.

Coastal Regions

Coastal regions embrace the ebb and flow of oceanic weather patterns, experiencing a unique dance with precipitation.

- **Consistency in Coastal Precipitation**: Coastal areas often enjoy more consistent rainfall, courtesy of the influence of oceanic weather patterns. This predictability enhances the reliability of

rainwater harvesting systems, offering a steady water source.
- **Tailoring Systems for Coastal Reliability**: Individuals in coastal regions can fine-tune their rainwater harvesting setups with a degree of confidence in the regularity of precipitation. Focus shifts toward optimizing storage and efficiency of use.

Inland Areas

In contrast, inland locations face a different climatic cadence. The variability in rainfall patterns necessitates strategic considerations for ensuring a reliable water supply.
- **Navigating Variable Rainfall:** Inland areas experience more variable rainfall, demanding a flexible approach to rainwater harvesting. Supplemental water storage and efficient collection systems become crucial for maintaining a steady water supply amid fluctuations.
- **Adaptability as a Key Virtue**: The adaptability of rainwater harvesting systems in inland areas becomes a virtue. Solutions that accommodate the unpredictability of rainfall allow for a more resilient water strategy.

Local Weather Patterns

Understanding the unique weather patterns of your region is paramount. Different climates, including arid, tropical, and temperate, present distinct challenges and opportunities for rainwater harvesting.

Arid Climates

Harvested water becomes a precious resource in arid regions, where rainfall is sporadic but potentially intense.
- **Efficient Storage in Arid Realms**: Rainwater harvesting in arid climates demands efficient storage and utilization practices. Each drop must be cherished, making water conservation an inherent part of the harvesting strategy.
- **Microscale Weather Patterns:** Even within arid regions, microscale weather patterns influence the effectiveness of rainwater harvesting systems. Understanding local nuances allows for more precise system design, acknowledging the intricacies of arid climates.

Tropical Climates

Tropical regions, blessed with heavy and frequent rainfall, pose challenges to managing excess water during intense storms.

- **Managing Tropical Abundance**: While the abundance of rainfall in tropical climates is advantageous, managing excess water during intense storms becomes a necessary consideration. Efficient drainage systems and storage solutions are essential.
- **Navigating Tropical Microclimates:** Tropical regions often host diverse microclimates. Urban areas within tropical zones experience different rainfall patterns compared to rural or coastal areas, requiring nuanced system designs.

Temperate Climates

Temperate climates exhibit seasonal variations in rainfall, requiring adaptability in rainwater harvesting systems.

- **Year-Round Adaptability:** Adapting rainwater harvesting systems to seasonal variations ensures water availability in temperate climates year-round. Flexibility becomes the key to harnessing nature's changing moods.
- **Monitoring Seasonal Shifts:** Recognizing shifts in temperature and precipitation patterns during different seasons allows for proactive adjustments in rainwater harvesting systems. Continuous monitoring guarantees year-round effectiveness.

Microclimates and Microscale Weather Patterns

Even within a relatively small geographical area, microclimates and microscale weather patterns can vary, adding a layer of complexity to rainwater harvesting.

Urban Heat Islands

Urban areas create localized heat islands that influence weather patterns, impacting rainfall and temperature.

- **Microclimates in Urban Jungles:** Urban heat islands introduce microclimates that diverge from broader regional patterns. Understanding these nuances allows for more precise system design, acknowledging the intricacies of city life.
- **Balancing Urban Development and Rainwater Harvesting:** Microclimates are intensified in urban settings where concrete

and asphalt dominate. Balancing the impermeability of urban surfaces with effective rainwater harvesting becomes crucial for sustainability.

Mountainous Terrain:

Mountainous regions experience orographic rainfall, influencing precipitation on windward and leeward sides.
https://www.pexels.com/photo/person-on-mountain-1647972/

Mountainous regions experience orographic rainfall, influencing precipitation on windward and leeward sides.

- **Navigating Mountain Dynamics:** In mountainous terrain, orographic rainfall leads to increased precipitation on windward sides and rain shadows on leeward sides. Strategic placement of harvesting systems considers these natural phenomena.
- **Harnessing Mountain Microclimates:** Microclimates within mountainous regions vary based on elevation, slope, and orientation. Understanding these intricacies aids in designing rainwater harvesting systems that align with the dynamic mountain environment.

Regulatory Considerations

Before finalizing your rainwater harvesting system, be aware of local regulations and guidelines. Regulatory considerations ensure a legal and sustainable approach to water collection.

Permitting and Regulations

Check if permits are required for rainwater harvesting systems. Some regions have regulations governing the size of storage tanks, runoff management, or water quality standards.

- **Navigating Legal Harmonies:** Understanding and complying with local permits and regulations ensures the legality and sustainability of your rainwater harvesting endeavors. Seek the necessary approvals to align your system with legal standards.
- **Educational Outreach on Regulatory Compliance:** Educate yourself and your community about the importance of adhering to regulations. Encourage awareness and compliance to foster a culture of legal and sustainable rainwater harvesting.

Community Guidelines

In communal living spaces or neighborhoods, adherence to community guidelines is essential. Collaborate with neighbors and local authorities to ensure your rainwater harvesting plans align with community standards.

- **Collective Responsibility:** Rainwater harvesting is not just an individual endeavor but a collective responsibility. Engage with your community to foster awareness and adherence to shared guidelines for sustainable water practices.
- **Sustainable Water Practices:** Work collaboratively with your community to establish guidelines that promote sustainable rainwater harvesting. Collective efforts enhance the effectiveness

and acceptance of rainwater harvesting practices within the community.

Orchestrating Harmony in Rainwater Harvesting

As you conclude your exploration of regional and climatic factors in rainwater harvesting, envision it as orchestrating a harmonious symphony with nature. Tailoring your system to the unique climate of your region transforms rainwater harvesting from a utilitarian task to a poetic interaction with the environment.

- **Strategic Adaptation:** Strategic adaptation is the hallmark of a well-designed rainwater harvesting system. Whether in arid deserts, tropical paradises, or temperate havens, your system's ability to adapt ensures harmony with nature.
- **Educational Outreach:** Share your experiences and knowledge about regional and climatic factors in rainwater harvesting with your community. Fostering awareness and understanding ensures a broader movement toward sustainable water practices.
- **Continuous Monitoring:** Nature is dynamic, and so should be your approach. Continuous monitoring of weather patterns, system efficiency, and local regulations ensures that your rainwater harvesting system remains in tune with the evolving nature of your surroundings.

In the end, rainwater harvesting is all about harmonizing with the rhythms of nature. As you design and implement your system, let the regional and climatic factors become the notes in a melody that celebrates the beauty and sustainability of water stewardship.

Maximizing capture efficiency requires a nuanced understanding of your property's unique characteristics and consideration of local climate and regulations. You'll find the key to running a harmonious and effective rainwater harvesting system in this intricate dance between practicality, aesthetics, and personal preferences.

Embrace the challenge of selecting the perfect spot, for in this choice, you unlock the potential to transform raindrops into a sustainable source of life for your home and the environment.

Chapter 4: Designing Your Harvesting System

The journey of rainwater harvesting takes a pivotal turn in this chapter, where you discover the art and science of designing a harvesting system. Transforming the transient dance of raindrops into a sustainable water source requires thoughtful considerations.

Basic Components of a Harvesting System

In the intricate relationship between the heavens and the Earth, rainwater emerges as a precious elixir, a gift bestowed upon humanity by nature. From the initial pattern of raindrops on catchment surfaces to the safeguarding embrace of storage tanks, each component plays a crucial role in harmonizing with the environment.

Catchment Surfaces

At the heart of every rainwater harvesting system lies the catchment surface. Roofs stand as the primary catchment surfaces, whether shingled, metal, or tiled. Each material contributes unique characteristics that influence the purity and volume of the harvested water.

Roof Materials and Purity

The choice of roofing materials plays a crucial role in determining the quality of the collected rainwater. Different materials bring distinct advantages and considerations to the table:

- **Metal Roofs**: Known for their durability, metal roofs, composed of corrosion-resistant materials like zinc or aluminum, minimize the introduction of contaminants. It makes them an excellent choice for maintaining water purity.

Known for their durability, metal roofs, composed of corrosion-resistant materials like zinc or aluminum, minimize the introduction of contaminants.
Wikideas1, CC0, via Wikimedia Commons:
https://commons.wikimedia.org/wiki/File:Standing_seam_metal_roof_low_pitch_roof-3.jpg

- **Asphalt Shingles:** Common in residential structures, asphalt shingles are cost-effective but may introduce small particles and contaminants into the harvested water. Implementing a first-flush diverter will mitigate these concerns.
- **Tile or Concrete Roofs:** These materials offer durability and aesthetic appeal. However, their surfaces contribute to water hardness or introduce minerals. Filtration systems will be necessary for quality control.

Expanding Catchment Beyond Roofs

While roofs serve as primary catchment surfaces, thinking beyond conventional structures opens avenues for innovation in rainwater harvesting. Consider exploring:

- **Awnings:** Extend the reach of your catchment system by strategically placing awnings. They complement roof catchments

and provide additional surfaces for rainwater collection.
- **Permeable Pavements**: Driveways and walkways made from permeable materials allow rainwater to penetrate the surface, contributing to the catchment. Incorporating these features will enhance the overall efficiency of your system.
- **Specially Designed Catchment Structures**: Innovative designs, such as catchment surfaces integrated into landscaping elements, will add both functionality and aesthetic appeal to your rainwater harvesting system.

Conveyance Systems

Once raindrops grace your catchment surface, the next act involves guiding them toward storage. Conveyance systems, comprising gutters and downspouts, conduct this liquid gold with efficiency and precision.

Gutter Systems

Well-designed gutter systems are the unsung heroes of rainwater harvesting. They ensure the smooth flow of water from the catchment surface to storage. Regular maintenance is essential to prevent blockages that compromise the efficiency of the entire conveyance system.

- **Material Considerations:** Choose gutter materials based on durability and compatibility with your catchment surface. Options include vinyl, aluminum, steel, and copper, each with its unique set of advantages.
- **Slope and Alignment:** Make sure that gutters are installed with a slight slope toward downspouts. Proper alignment prevents water stagnation and facilitates efficient drainage.

Downspouts and Diverter Systems

Downspouts act as conductors, guiding water from gutters to storage, while diverter systems enhance efficiency by preventing initial runoff, laden with debris and contaminants, from reaching storage directly.

- **Diverter Types:** First-flush diverters are crucial components. They redirect the initial runoff, which contains pollutants washed from the catchment surface, ensuring that only cleaner water enters the storage system.
- **Regular Maintenance:** Inspect and clean downspouts and diverter systems regularly to prevent clogs and blockages. This maintenance practice preserves the integrity and efficiency of your rainwater conveyance.

Filters

Before rainwater cascades into storage, it undergoes a refining process through filters that sift through impurities.

Mesh Screens

Basic mesh screens effectively capture larger debris, such as leaves and twigs, preventing them from entering the storage system. Regular cleaning is crucial to prevent clogging and maintain optimal filtration efficiency.

- **Maintenance Routine**: Include regular checks and cleaning of mesh screens in your rainwater harvesting maintenance routine. This simple step goes a long way toward preserving the functionality of your filtration system.
- **Cartridge Filters:** Cartridge filters become indispensable for finer filtration, especially in systems designed for potable water. These filters come in various micron ratings, allowing you to tailor filtration to the specific contaminants present in your region.
- **Micron Ratings**: Choose cartridge filters with appropriate micron ratings based on the quality of the harvested water and the contaminants you aim to remove. This precision ensures the purity of your collected rainwater.

Storage Tanks

The final destination for harvested rainwater is the storage tank. Tanks come in various materials, sizes, and shapes, each tailored to specific needs and space constraints.

Tank Materials

The choice of tank material influences durability, cost, and water quality. Consider the following options based on your preferences and specific requirements:

- **Polyethylene Tanks**: Lightweight and cost-effective polyethylene tanks are suitable for above-ground installations. They are resistant to corrosion and provide a practical solution for many applications.
- **Concrete Tanks:** Durable and suitable for underground installations, concrete tanks offer longevity and stability. However, proper sealing is essential to prevent minerals from leaching into the stored water.

- **Underground Cisterns:** Concealed beneath the ground, these tanks provide space-saving solutions. The choice of materials remains crucial to prevent contamination and ensure the purity of stored water.

Sizing Considerations

Calculating the ideal tank size involves considerations such as catchment area, average rainfall, and intended use. Oversized tanks ensure ample reserves for drier periods, offering a buffer against water scarcity.

- **Catchment Potential:** Evaluate the catchment potential of your surfaces, including roofs and additional catchment structures. This calculation forms the basis for determining the required storage capacity.
- **Average Rainfall:** Consider the average annual rainfall in your region. This data helps estimate the potential volume of harvested water, aiding in the selection of an appropriately sized storage tank.
- **Intended Use:** Define the purpose of your harvested water, whether for irrigation, household use, or potable water. Each use dictates the required volume, influencing the sizing of your storage tank.

Sustainability

Each component is fine-tuned by thoughtful design, from the catchment surfaces to the storage tanks.

- **Balancing Act:** Achieving a balance between functionality, aesthetics, and sustainability is key. Consider how each component contributes not only to the efficiency of your system but also to its overall impact on the environment.
- **Stewardship Responsibility:** Embrace the role of a responsible steward of water resources. The choices made in designing and implementing your rainwater harvesting system ripple through the broader ecosystem, reflecting a commitment to sustainability.
- **Continuous Care:** As you start to harness the rain's bounty, remember that continuous care is essential. Regular maintenance of components ensures the longevity and efficiency of your system, ensuring a seamless continuation of the liquid magic.

In the next section, you'll learn about the intricate process of designing a rainwater harvesting system tailored to your specific needs and the unique characteristics of your environment.

Considerations Based on Intended Use

Rainwater harvesting transforms the art of utilizing rain into a versatile and sustainable resource. Each factor is carefully tuned to the intended use, whether nourishing the earth through irrigation, elevating daily household chores, or satisfying the thirst with potable water. It's time to dive deeper into each component, exploring the intricacies and considerations that make rainwater harvesting a personalized and sustainable endeavor.

Drip Irrigation Systems

Irrigation is a dance between water and soil. Here, precision and efficiency take center stage, with drip irrigation systems orchestrating a symphony of water droplets to nourish the earth. For those cultivating the earth, harvested rainwater becomes a lifeline for crops and greenery. Designing a system for irrigation requires considerations beyond purity, emphasizing volume and distribution efficiency. It's time to explore how this movement unfolds, ensuring that every drop fulfills its purpose in the grand composition of rainwater harvesting.

- **Efficient Water Distribution:** Drip irrigation systems minimize wastage with their ability to deliver water precisely to the plants' root zones. Coupled with appropriate filters, these systems ensure the efficient distribution of water, optimizing its use for agricultural purposes.
- **Sizing for Specific Needs:** Calculating the irrigation demand involves a nuanced understanding of the types of plants, soil characteristics, and local climate. A well-designed system aligns the availability of rainwater with the specific needs of the green space, promoting sustainable agricultural practices.

Household Filtration Systems

In a household context, rainwater elevates mundane chores to sustainable practices. Tailoring the system for household use involves addressing water quality and distribution for various domestic needs.

- **Enhancing Water Quality:** Household filtration systems play a pivotal role in ensuring water quality for domestic use. Using filters tailored to remove specific contaminants, such as sediment filters, activated carbon, or UV purification, enhances the purity of rainwater for everyday tasks.

- **Seamless Integration:** Designing the system to seamlessly integrate with household plumbing is crucial. Incorporating pressure pumps and distribution networks ensures a reliable supply for everyday tasks, transforming rainwater into a sustainable resource for daily living.

Drinking Water

For those venturing into potable rainwater, the design takes on a heightened level of precision. Rigorous filtration, disinfection, and compliance with health standards become paramount.

- **Multi-Barrier Filtration**: Employ advanced filtration systems with a multi-barrier approach. This may include sediment filtration, activated carbon, UV treatment, and, in some cases, reverse osmosis. Each layer contributes to the overall purity of harvested rainwater.
- **Continuous Monitoring:** Regularly testing the harvested water for contaminants is essential. Compliance with local health regulations ensures the potability of the water harvested, transforming rain into a safe and sustainable source of drinking water.

Scalability and Adaptability in Design

An effective rainwater harvesting system isn't just a static structure. It's a dynamic entity capable of growth. Scalability ensures that as needs evolve, the system expands to accommodate increased demand.

- **Oversizing Components**: Opt for oversized catchment surfaces and storage tanks. This provides a buffer for future expansion without requiring substantial redesign, allowing the system to grow in harmony with your evolving requirements.
- **Modular Conveyance Systems**: Design gutters and downspouts in a modular fashion. This allows for easy additions or modifications as catchment areas expand, facilitating seamless scalability without disrupting the existing structure.

Adaptability

Nature is dynamic, and so is the environment around your rainwater harvesting system. Designing with adaptability in mind ensures resilience against unforeseen changes and challenges.

- **Flexibility in Filtration**: Choose filter systems with modular components. This facilitates adjustments based on changes in

water quality or the introduction of new contaminants, ensuring that the system can adapt to evolving conditions.
- **Weather-Responsive Controls**: Integrate weather-responsive controls for irrigation systems. This ensures adjustments based on rainfall forecasts, preventing overwatering during rainy periods. The adaptability to changing weather patterns makes the system responsive and efficient.

In the end, rainwater harvesting is all about harmonizing with the rhythms of nature. As you design and implement your system, let the intended use guide the composition, creating a masterpiece that transforms rain into a versatile and sustainable resource.

The Design Process

Rainwater harvesting is not just a pragmatic endeavor. It's a combination of thoughtful design, meticulous planning, and harmonious integration with nature. As you delve into the intricacies of the design process, envision it as composing a piece that resonates with the unique cadence of your environment. Here are the steps of this creative process, where each decision is a note in the melody of sustainability.

Step 1: Assessment of Catchment Potential

Evaluate Roof Characteristics

Assessing your roof's type, size, and material sets the architectural tone for your rainwater harvesting system. Each characteristic influences catchment potential and water quality.

- **Type:** Each type presents unique challenges and opportunities, from flat roofs to sloped designs. Assess how the architectural nuances impact water runoff and collection efficiency.
- **Size:** The size of your roof is a crucial factor in determining the catchment area. Larger roofs offer more potential for water collection but also demand careful considerations in system design.
- **Material:** The material of your roof goes beyond aesthetics. Different materials can introduce contaminants or enhance water purity. Consider corrosion-resistant metals or durable synthetic materials for optimal results.

Explore Additional Catchment Surfaces

Beyond the roof, additional catchment surfaces contribute to the richness of your water-harvesting efforts. A comprehensive assessment ensures optimal utilization of available surfaces.

- **Walls and Awnings**: Vertical surfaces like walls and awnings supplement your roof's catchment potential. Assess their contribution to overall water collection and factor them into your design.
- **Permeable Surfaces**: Evaluate permeable surfaces like driveways or courtyards. While these may not contribute directly to water catchment, understanding their role in water flow aids in efficient system design.

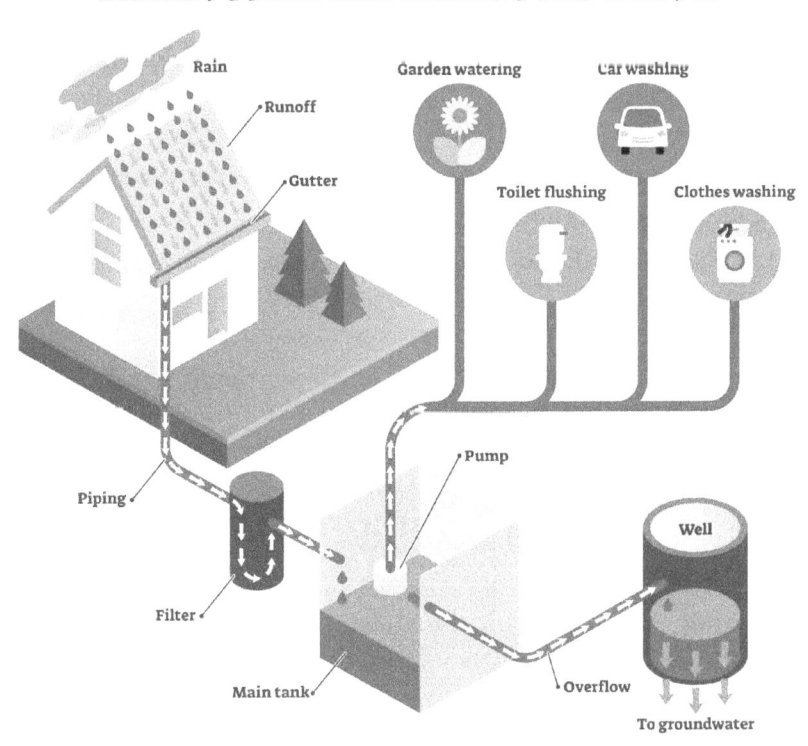

RAINWATER HARVESTING

Beyond the roof, additional catchment surfaces contribute to the richness of your water-harvesting efforts.

Step 2: Calculating Water Demand

Determine Intended Use

Clearly defining the purpose of harvested water makes your goals clearer and your efforts focused. Each use dictates the required volume and quality for irrigation, household use, or potable water.

- **Irrigation:** If your focus is on irrigation, the demand may vary based on the types of plants and the size of the landscaped area. Understanding the specific water needs of your plants guides the design process.
- **Household Use:** For household use, consider daily activities like cooking, cleaning, and bathing. Clearly outlining the intended use ensures that your system aligns with practical water needs.
- **Potable Water:** If your goal is to harvest water for drinking, the highest purity standards are essential. The design must incorporate advanced filtration and purification components.

Calculate Demand

Based on the intended use, calculate the daily and seasonal water demand. This serves as the foundation for sizing components and designing the conveyance and storage systems.

- **Daily Demand:** Consider the daily water requirements for your chosen purpose. This includes understanding peak usage times and designing the system to meet these demands.
- **Seasonal Variations:** Recognize how water needs may vary throughout the seasons. Designing for seasonal fluctuations ensures a reliable supply of harvested water year-round.

Step 3: Selecting Appropriate Components

Roof and Catchment Design

Choosing roof materials and designing catchment surfaces is where aesthetics, durability, and efficiency come together harmoniously.

- **Aesthetics:** The visual appeal of your roof and catchment surfaces is integral to the overall design. Consider materials and designs that complement the architectural style of your property.
- **Durability:** Longevity is a key consideration. Select materials that withstand weathering and environmental factors, ensuring the

sustained efficiency of your rainwater harvesting system.
- **Efficiency**: Striking the right balance between aesthetics and durability guarantees that your roof and catchment surfaces efficiently channel water to the collection system.

Conveyance Systems

Selecting gutter and downspout systems impacts the channels through which your water flows. Incorporate first-flush diverters and establish regular maintenance protocols.

- **Gutter Systems:** Choose gutter systems suitable for the catchment area. Consider factors like material, size, and shape to optimize water flow. Regular cleaning and maintenance prevent blockages.
- **Downspouts**: Efficient downspouts guide water downward from the roof to storage tanks. Position them strategically to maximize water capture and minimize runoff.
- **First-Flush Diverters:** Integrate first-flush diverters to minimize the initial runoff that may carry contaminants. This enhances the overall quality of harvested water.

Filtration Solutions

Choose appropriate filtration components based on water quality goals. Mesh screens, cartridge filters, or advanced purification systems ensure the desired water purity.

- **Mesh Screens:** These act as the first line of defense, preventing larger debris from entering the system. Regular cleaning maintains its effectiveness.
- **Cartridge Filters**: Mid-level filtration components that capture smaller particles. Choose cartridges based on your water quality objectives.
- **Advanced Purification**: Consider advanced purification systems like UV filters or reverse osmosis for potable water purposes. These ensure the highest level of water purity.

Storage Tanks

Consider tank materials and sizing based on catchment potential and water demand. Factor in scalability for future expansion, allowing your system to grow with changing needs.

- **Material Selection:** Choose materials that are durable, non-toxic, and corrosion-resistant. The most common include polyethylene,

fiberglass, and concrete.
- **Sizing Considerations:** Your storage tank size should align with catchment potential and water demand. Calculate the necessary storage capacity to ensure a reliable water supply.
- **Scalability:** Opt for oversized storage tanks and consider modular tank designs. This accommodates future expansion without the need for significant redesign.

The size of your storage tanks should align with both catchment potential and water demand.
SuSanA Secretariat, CC BY 2.0 <https://creativecommons.org/licenses/by/2.0>, via Wikimedia Commons: https://commons.wikimedia.org/wiki/File:Variety_of_water_storage_tanks_and_rainwater_harvesting_equipment_(4481564110).jpg

Step 4: System Integration and Distribution Planning

Integration with Existing Structures

Seamlessly integrate the rainwater harvesting system with existing structures. This includes plumbing for household use, irrigation networks, and potential expansion points.

- **Plumbing Integration:** Connect the rainwater harvesting system to existing plumbing for household use. Make sure that the harvested water seamlessly integrates with conventional water sources.

- **Irrigation Networks**: If the system is used for irrigation, plan for the integration of the rainwater supply with existing or new irrigation networks. Distribute water efficiently to landscaped areas.

Distribution Planning

Plan for efficient water distribution based on intended use. This may involve pressure pumps, drip irrigation networks, or household plumbing adjustments.

- **Pressure Pumps:** If needed, incorporate pressure pumps to ensure adequate water pressure for household use or irrigation. Proper distribution relies on maintaining consistent pressure.
- **Drip Irrigation Networks:** For irrigation purposes, design drip irrigation networks that deliver water directly to the base of plants. This conserves water and ensures targeted hydration.
- **Household Plumbing Adjustments**: If integrating with household use, plan for adjustments in plumbing to facilitate the seamless incorporation of harvested water into daily activities.

Step 5: Adaptability and Scalability Features

Modular Conveyance Additions

Design gutters and downspouts in a modular fashion. This allows for easy additions or modifications as catchment areas expand.

- **Modular Design:** Create a gutter and downspout system that can be easily extended or modified. This ensures adaptability as you expand your rainwater harvesting reach.
- **Future Catchment Areas**: Anticipate potential future catchment areas and design the system to accommodate these additions. This future-oriented approach prevents the need for significant overhauls.

Scalable Storage

Opt for oversized storage tanks and consider modular tank designs. This accommodates future expansion without the need for significant redesign.

- **Oversized Tanks**: Select storage tanks with a capacity that exceeds your current demand. This surplus capacity prepares your system for increased water needs in the future.

- **Modular Tank Designs:** Choose tank designs that allow for the easy addition of new modules. This scalable approach guarantees that your storage capacity can evolve with changing requirements.

Adaptable Filtration Systems

Choose filtration systems with modular components. This facilitates adjustments based on changes in water quality or the introduction of new contaminants.

- **Modular Filtration:** Choose filtration systems with interchangeable components. This allows you to upgrade or modify the system to address evolving water quality concerns.
- **Contaminant-Specific Filtration:** If the water source changes, such as increased sedimentation, choose filtration components specifically targeting the identified contaminants. This ensures continued water purity.

Step 6: Weather-Responsive Controls (Optional)

Implementing Smart Controls

For irrigation systems, consider weather-responsive controls. These systems adjust watering schedules based on real-time weather data, preventing overwatering during rainy periods.

- **Smart Controllers:** Incorporate weather-responsive controllers into your irrigation system. These controllers use real-time weather data to adjust watering schedules, optimizing water use.
- **Rain Sensors:** Integrate rain sensors that automatically suspend irrigation during rainfall. This ensures that harvested rainwater is not wasted and promotes water conservation.

Monitoring and Adjustment Protocols

Establish protocols to monitor system performance. Regular checks, especially after significant weather events, ensure optimal function and allow for adjustments as needed.

- **Routine Checks:** Schedule routine checks of the entire rainwater harvesting system. Inspect gutters, downspouts, filtration systems, and storage tanks to promptly identify and address any issues.
- **Post-Weather Event Checks:** After significant weather events, conduct thorough inspections. Heavy rainfall or storms may impact system components, and proactive checks prevent

potential problems.

- **Adjustment Protocols:** Develop clear protocols for making adjustments to the system. Systematic adjustments maintain system efficiency, whether adapting to changing water quality or expanding the catchment area.

A Blueprint for Sustainable Water

Your rainwater harvesting system is a functional setup and a blueprint for sustainable water stewardship. Each decision, from catchment selection to storage tank sizing, becomes a stroke in the canvas of environmental responsibility.

- **Holistic Water Management:** Designing a rainwater harvesting system is a holistic approach to water management. It's a conscious choice to nurture nature's gift responsibly.
- **Adaptability for the Future:** As you face an ever-changing world, your rainwater harvesting system becomes a beacon of adaptability. Scalability and flexibility ensure it meets the challenges and opportunities of the future.
- **Educational Outreach:** Share your design insights with your community. Foster awareness and understanding of rainwater harvesting as a sustainable practice. Encourage others to embark on the journey of designing their systems.

The intricate design process of rainwater harvesting is a composition that harmonizes with the natural rhythms of your surroundings. Each decision, from selecting roof materials to integrating distribution networks, contributes to the seamless flow of your water-harvesting melody. Let your rainwater harvesting system be a testament to the artistry of sustainable living, where every drop is a note in the symphony of water stewardship.

Chapter 5: Storage Systems – Barrels, Gutters, and Tanks

Rainwater harvesting is a sustainable practice that conserves water and provides an eco-friendly alternative for various purposes. Choosing the right storage solution is paramount to the success of your rainwater harvesting venture. From the simplicity of gutters and barrels to the substantial presence of tanks, understanding the diverse storage options will empower you to make informed decisions that align with your needs, budget, and environmental considerations.

Short-Term Storage: Gutters and Barrels

Rainwater harvesting stands at the forefront of sustainable practices, offering a conscientious approach to water conservation. Gutters and barrels are two critical components of rainwater harvesting's initial stages. These short-term storage solutions play a pivotal role in collecting rainwater efficiently, providing an immediate and accessible source for a multitude of purposes. From materials and capacities to advantages and drawbacks, it's time to navigate the intricacies of these essential components.

Gutters

Gutters, the silent architects of rainwater harvesting, form the inaugural phase of this sustainable practice.

Stilfehler, CC BY-SA 4.0 <https://creativecommons.org/licenses/by-sa/4.0>, via Wikimedia Commons: https://commons.wikimedia.org/wiki/File:Upstate_New_York_Seamless_Aluminum_Gutters_02.jpg

Gutters, the silent architects of rainwater harvesting, form the inaugural phase of this sustainable practice. As the first line of defense, gutters are crafted from materials such as aluminum, steel, or PVC. These unassuming channels, elegantly positioned along the roofline, play a crucial role in directing rainwater into downspouts, initiating the journey of collected water. Here's an exploration of the diverse materials, capacities, advantages, and drawbacks of gutters, unraveling the simplicity and effectiveness they bring to short-term rainwater storage.

Materials

Often crafted from materials like aluminum, steel, or PVC, gutters serve as the essential conduits that collect rainwater along the roofline, channeling it into downspouts. Each material brings its unique set of advantages to the rainwater harvesting ecosystem.

- **Aluminum:** Renowned for their lightweight nature and corrosion resistance, aluminum gutters are a preferred choice. Their durability and ease of handling make them practical and efficient

solutions for homeowners seeking a reliable rainwater harvesting system.
- **Steel**: Recognized for robustness, steel gutters are a durable option. However, the susceptibility to rust requires regular maintenance to prevent deterioration over time, making them suitable for those willing to invest time in upkeep.
- **PVC:** As a cost-effective alternative, PVC gutters are resistant to both corrosion and rust. The versatility of PVC gutters is a standout feature, allowing them to adapt seamlessly to various roof types and shapes.

Capacities

The efficacy of gutters is closely tied to their size and the regional rainfall patterns. Regular maintenance, involving tasks such as clearing debris, is imperative to ensure optimal water flow and prevent overflow during heavy rainfall.

- **Size Variation**: Gutters come in various sizes, accommodating the diverse needs of different properties. Larger gutters handle more significant volumes of water, making them suitable for regions with higher rainfall.
- **Rainfall Considerations:** The capacity of gutters is directly influenced by the amount of rainfall in a specific area. Understanding regional rainfall patterns is crucial to determining the appropriate gutter size to collect and manage rainwater effectively.
- **Integration with Downspouts**: The seamless integration with downspouts also influences the capacity. Properly designed systems ensure water efficiently travels from the gutters to the downspouts, preventing overflow and maximizing storage.
- **Maintenance Impact**: Regular maintenance, such as clearing leaves and debris, is paramount for ensuring optimal water flow. Well-maintained gutters effectively manage higher capacities without the risk of clogs or overflowing.

Advantages

Gutters boast several advantages, making them an attractive choice for short-term rainwater storage:

- **Cost-Effective:** Gutters present a budget-friendly option, democratizing rainwater harvesting by making it accessible to a

Gutters

Gutters, the silent architects of rainwater harvesting, form the inaugural phase of this sustainable practice.
Stilfehler, CC BY-SA 4.0 <https://creativecommons.org/licenses/by-sa/4.0>, via Wikimedia Commons: https://commons.wikimedia.org/wiki/File:Upstate_New_York_Seamless_Aluminum_Gutters_02.jpg

Gutters, the silent architects of rainwater harvesting, form the inaugural phase of this sustainable practice. As the first line of defense, gutters are crafted from materials such as aluminum, steel, or PVC. These unassuming channels, elegantly positioned along the roofline, play a crucial role in directing rainwater into downspouts, initiating the journey of collected water. Here's an exploration of the diverse materials, capacities, advantages, and drawbacks of gutters, unraveling the simplicity and effectiveness they bring to short-term rainwater storage.

Materials

Often crafted from materials like aluminum, steel, or PVC, gutters serve as the essential conduits that collect rainwater along the roofline, channeling it into downspouts. Each material brings its unique set of advantages to the rainwater harvesting ecosystem.

- **Aluminum:** Renowned for their lightweight nature and corrosion resistance, aluminum gutters are a preferred choice. Their durability and ease of handling make them practical and efficient

solutions for homeowners seeking a reliable rainwater harvesting system.
- **Steel**: Recognized for robustness, steel gutters are a durable option. However, the susceptibility to rust requires regular maintenance to prevent deterioration over time, making them suitable for those willing to invest time in upkeep.
- **PVC:** As a cost-effective alternative, PVC gutters are resistant to both corrosion and rust. The versatility of PVC gutters is a standout feature, allowing them to adapt seamlessly to various roof types and shapes.

Capacities

The efficacy of gutters is closely tied to their size and the regional rainfall patterns. Regular maintenance, involving tasks such as clearing debris, is imperative to ensure optimal water flow and prevent overflow during heavy rainfall.

- **Size Variation**: Gutters come in various sizes, accommodating the diverse needs of different properties. Larger gutters handle more significant volumes of water, making them suitable for regions with higher rainfall.
- **Rainfall Considerations:** The capacity of gutters is directly influenced by the amount of rainfall in a specific area. Understanding regional rainfall patterns is crucial to determining the appropriate gutter size to collect and manage rainwater effectively.
- **Integration with Downspouts**: The seamless integration with downspouts also influences the capacity. Properly designed systems ensure water efficiently travels from the gutters to the downspouts, preventing overflow and maximizing storage.
- **Maintenance Impact**: Regular maintenance, such as clearing leaves and debris, is paramount for ensuring optimal water flow. Well-maintained gutters effectively manage higher capacities without the risk of clogs or overflowing.

Advantages

Gutters boast several advantages, making them an attractive choice for short-term rainwater storage:

- **Cost-Effective:** Gutters present a budget-friendly option, democratizing rainwater harvesting by making it accessible to a wide spectrum of homeowners.

- **Ease of Installation:** The simplicity of gutter systems translates into easy installation, often making them a popular choice for those who enjoy engaging in DIY projects around their homes.
- **Versatility**: The adaptability of gutters to different roof types and shapes adds to their appeal. This versatility makes them suitable for a variety of architectural designs.

Drawbacks

- **Limited Storage Capacity:** Gutters are not designed for extensive water storage, rendering them more suitable for immediate use rather than long-term storage solutions.
- **Maintenance Requirements**: Frequent cleaning is necessary to prevent clogs and overflow, demanding regular homeowner attention. While the maintenance is straightforward, it's an ongoing commitment.

Barrels

As you move along the rainwater harvesting journey, barrels step onto the stage, providing an elegant enhancement to short-term storage solutions. Positioned strategically under downspouts, these unassuming containers elevate the rainwater collection, offering functionality and an aesthetic touch to the process. In this section, you'll discover rain barrels' various materials, capacities, advantages, and drawbacks, unraveling their role in enhancing short-term rainwater storage.

Materials

Rain barrels strategically positioned under downspouts complement gutters in collecting rainwater. These barrels come in various materials, each with unique merits, adding a layer of customization to the rainwater harvesting experience.

- **Plastic:** Lightweight and corrosion-resistant, plastic barrels are a popular and practical choice. Their ease of handling and suitability for various climates make them a go-to option for many homeowners.
- **Wood:** For those looking for a touch of rustic aesthetics, wooden barrels fit the bill. While they add an attractive element to the garden landscape, they require more maintenance to preserve their charm.

- **Metal:** Known for their durability, metal barrels are sturdy and chosen for their longevity. The trade-off is their susceptibility to rust, requiring you to weigh the benefits against potential maintenance needs.

Capacities

Rain barrels offer a range of capacities, typically 50 to 100 gallons. This variability allows you to select a size that aligns with your water needs and the available space on your property.

- **Size Options:** Rain barrels come in various sizes to cater to different usage requirements. Smaller barrels are suitable for limited-space areas, while larger ones accommodate higher water demands.
- **Modularity:** You can install multiple barrels in a modular fashion, creating a collective storage system with increased capacity. This modularity provides flexibility in adapting to changing water consumption patterns.
- **Customization:** Some barrels are designed with customizable capacities, allowing you to choose the size that best fits your specific needs. This customization ensures that the rain barrel aligns seamlessly with the property's requirements.
- **Overflow Prevention Features:** Many rain barrels incorporate features like overflow valves or outlets to manage higher capacities effectively. These mechanisms ensure excess water is directed away from the barrel, preventing overflow and potential water wastage.

Advantages

- **Affordability:** Rain barrels are cost-effective, align with budget considerations, and make rainwater harvesting accessible to a broad audience.
- **Easy Installation:** Like gutters, rain barrels are relatively easy to install, often making them popular for DIY projects. This simplicity adds to their appeal, especially for those with a hands-on approach to home improvement.
- **Immediate Access**: Rain barrels offer immediate access to harvested rainwater, facilitating activities like watering plants or washing outdoor surfaces without delay.

Drawbacks
- **Limited Storage Capacity:** Like gutters, rain barrels are not designed for extensive storage. They are best suited for short-term use, emphasizing immediate accessibility over prolonged storage needs.
- **Maintenance Requirements:** Regular cleaning and filtering are necessary to ensure water quality and prevent issues such as mosquito breeding. While maintenance is crucial, it is a manageable aspect for those committed to reaping the benefits of rainwater harvesting.

Understanding the materials, capacities, advantages, and drawbacks of gutters and barrels is pivotal for making informed decisions. Homeowners looking to embark on a rainwater harvesting journey can blend the efficiency of gutters with the accessibility of barrels to create a well-rounded and sustainable system. By embracing these short-term solutions, you contribute to your water conservation efforts and the broader movement toward environmentally conscious living.

Making an informed decision about short-term rainwater harvesting solutions requires thoughtful consideration of individual needs, property characteristics, and commitment to maintenance. The synergy of gutters and barrels provides a balanced approach, offering efficiency and accessibility to homeowners seeking to integrate sustainability into their daily lives.

Long-Term Storage: Tanks

In the quest for sustainable water management, long-term rainwater harvesting solutions become paramount, especially for those facing infrequent rainfall or high water demand scenarios. Tanks emerge as stalwart players in this arena, offering substantial storage capacities to meet the needs of residential, commercial, and agricultural applications.

Materials

Tanks, the cornerstone of long-term rainwater harvesting, are crafted from a variety of materials, each presenting a unique set of characteristics.
- **Polyethylene Tanks:** Lightweight and resistant to corrosion, polyethylene tanks offer a practical solution for those seeking durability without the burden of excessive weight. Their versatility extends to above-ground installations, making them accessible for

various applications.
- **Fiberglass Tanks**: Renowned for their durability, fiberglass tanks are a robust choice, particularly suitable for underground installation. This feature preserves property aesthetics and optimizes space usage. Fiberglass tanks are resistant to corrosion, making them a reliable long-term option.
- **Concrete Tanks**: Robust and sturdy concrete tanks are known for their durability. However, their weight is a limiting factor, and they are typically employed in scenarios where above-ground installation is feasible. Concrete tanks provide a solid and long-lasting solution for extensive rainwater storage needs.
- **Steel Tanks:** Sturdy and capable of withstanding external pressures, steel tanks are a common choice for above-ground installations. However, they are prone to rust, necessitating careful consideration of maintenance practices to ensure their longevity.

Capacities

The allure of tanks lies in their ability to cater to a broad spectrum of water storage needs, from modest residential applications to large-scale commercial and agricultural requirements.

- **Range of Capacities:** Tanks offer a diverse range of capacities, accommodating the specific demands of various users. From a few hundred gallons to several thousand, the flexibility in size ensures that individuals and businesses can tailor their rainwater harvesting systems to their unique requirements.
- **Residential Applications:** Smaller tank capacities are often suitable for residential applications, providing homeowners with a reliable and sustainable water source for domestic use, landscaping, and other household needs.
- **Commercial and Agricultural Needs**: Larger tank capacities find their niche in commercial and agricultural settings where water demand is more substantial. Tanks are integral to ensuring a consistent and sufficient water supply for crops, livestock, and industrial processes.

Advantages

Tanks bring many advantages to the table, making them indispensable for those seeking robust, long-term rainwater harvesting solutions.

- **Substantial Storage Capacity**: The primary strength of tanks lies in their ability to store significant volumes of water, making them ideal for regions with infrequent rainfall or areas facing high water demand. This feature gives you a reliable and consistent water supply even during dry spells.
- **Customization for Underground Installation:** Tanks can be customized for underground installation. They're a particularly valuable option for those aiming to preserve above-ground space or maintain property aesthetics. This underground configuration enhances space efficiency and protects the tanks from external elements.
- **Versatility in Applications**: Tanks cater to a wide array of applications, from residential water conservation to large-scale agricultural and commercial operations. Their adaptability positions them as versatile solutions for diverse water storage needs.

Drawbacks

While tanks offer substantial benefits, it's essential to acknowledge the considerations and challenges associated with implementing them.

- **Higher Upfront Costs:** Compared to short-term solutions like gutters and barrels, tanks come with higher upfront costs. The investment required to purchase and install tanks is a significant consideration for those managing budget constraints.
- **Professional Installation Required**: Installing tanks, especially in customized or underground configurations, often requires professional assistance. It adds to the overall costs and underscores the importance of ensuring the installation is carried out precisely.
- **Regular Maintenance is Crucial:** Regular maintenance is imperative to ensure tank longevity and optimal functionality. It includes checking for corrosion, cleaning, and addressing any potential issues promptly. Ignoring maintenance will lead to deterioration and reduce the lifespan of the system.

Implementing Tanks

As you navigate long-term rainwater harvesting, implementing tanks emerges as a pivotal step toward building water resilience. Understanding tanks' materials, capacities, advantages, and drawbacks equips individuals and businesses to make informed decisions aligning with their specific needs and circumstances.

- **Material Selection:** Choosing the right material for a tank involves a delicate balance between durability, weight considerations, and the intended application. Polyethylene tanks, with their lightweight nature and corrosion resistance, are ideal for above-ground residential installations. Fiberglass tanks, with their durability and suitability for underground use, offer a discreet solution that doesn't compromise property aesthetics. Concrete tanks, though heavier, provide robustness for various applications, while steel tanks, which are also sturdy, require diligent maintenance to combat rust.
- **Calculating Capacities**: Determining the appropriate tank capacity depends on accurately assessing water demand. Residential users may find smaller capacities sufficient for daily needs, while those engaged in agriculture or commercial activities require larger tanks for a consistent and reliable water supply. Understanding regional rainfall patterns and the frequency of dry spells aids in fine-tuning the capacity to match actual needs.
- **Advantages for Varied Applications**: The versatility of tanks shines in their ability to cater to a spectrum of applications. In a residential setting, tanks are a sustainable water source for daily activities, landscape irrigation, and emergency preparedness. For commercial enterprises, tanks provide a reliable supply for industrial processes, reducing dependence on external water sources. In agriculture, where water is a lifeline for crops and livestock, tanks guarantee a consistent supply, contributing to sustainability and productivity.
- **Mitigating Drawbacks**: While the drawbacks associated with tanks are notable, proactive planning mitigates potential challenges. Addressing the higher upfront costs involves considering the long-term benefits and return on investment that tanks bring. Seeking professional assistance during installation ensures the system is set up correctly, maximizing its efficiency and lifespan. Finally, regular maintenance should be viewed as a proactive investment rather than a reactive necessity, safeguarding the longevity and functionality of the rainwater harvesting system.

Using tanks in a rainwater harvesting system is not just a practical choice. It's a commitment to sustainable water practices. As individuals, communities, and businesses strive to reduce their environmental footprint, the role of tanks in water conservation becomes increasingly pivotal. The investment in tanks transcends the mere procurement of a storage system. It symbolizes a dedication to responsible water management and a proactive stance in securing water resources for the future.

As you integrate tanks into your rainwater harvesting endeavors, you contribute to personal water security and the broader movement promoting water resilience. Adopting long-term solutions like tanks represents a collective step towards a more sustainable and water-conscious future, where every drop is valued, conserved, and utilized precisely.

Maintenance and Integration

A well-maintained rainwater harvesting system is not just a reservoir for environmental consciousness but an investment in efficiency and longevity. Whether your chosen storage option is barrels, tanks, or a combination of both, routine maintenance is the key to its effectiveness. It's time to dive into the specifics to ensure your system runs like a finely tuned instrument.

Gutters

1. **Regular Cleaning:** The first line of defense in rainwater harvesting is your gutters. These conduits direct rain from your roof into the storage system. Regular cleaning is paramount to prevent clogs and ensure a smooth flow of water. The flow is restricted if clogged, just like cholesterol blocking blood vessels. Regular cleaning keeps the path clear and fluid.

2. **Inspection and Repair:** Take a closer look at your gutters periodically. Are they sagging or damaged? Repair any issues promptly to maintain their structural integrity. Gutters, much like a well-tuned instrument, need occasional adjustments. Tighten loose screws or replace damaged sections to keep everything in harmony.

3. **Leak Checks:** The connections between gutters and downspouts are potential leak points. Regularly inspect these junctions and fix any leaks promptly. A small leak will disrupt the entire

performance.

Barrels

1. **Interior Cleanliness:** The interior of your barrels is a breeding ground for potential issues, especially algae growth. Regular cleaning prevents these unwanted green guests.
2. **Crack and Leak Inspection:** Barrels, like any container, are susceptible to cracks and leaks. Periodically inspect them and promptly repair any damages. Just as a musician checks their instrument for cracks or warps, you must inspect your barrels. A small crack may seem insignificant, but it leads to a loss of the precious liquid.
3. **Securing the Lid:** The lid of your barrel is its first line of defense against debris and contamination. Ensure it's securely in place to maintain the purity of your harvested rainwater.

Tanks

1. **Structural Inspections:** Being larger structures, tanks require periodic inspections for leaks or structural damage. Catching these issues early on prevents more extensive problems. A tank with structural damage is a building with a compromised foundation. Regular inspections ensure everything stands strong.
2. **Interior Cleanliness:** Sediment buildup in tanks reduces their efficiency. Regularly clean the tank's interior to prevent this, ensuring a clear and unobstructed space for water storage. Routine cleaning keeps everything in top condition.
3. **Filtration System Monitoring:** If your tank has a filtration system, monitor it regularly for optimal performance. Clean or replace filters as needed. Filters in a rainwater harvesting system are like the strings on a guitar. Regular tuning (cleaning or replacement) ensures they produce the desired melody.

Harmonizing Functionality and Aesthetics

1. **Strategic Placement:** Consider the strategic placement of your storage systems. Optimize water collection while minimizing the visual impact on your property. Placing tanks strategically is like arranging furniture in a room. It must be functional (providing seating or storage) while not overpowering the visual space.
2. **Aesthetics:** Choose storage options that complement your property's aesthetic. Whether it's the sleek lines of a tank or the

rustic charm of barrels, make sure it aligns with the visual theme. It should enhance, not detract, from the overall aesthetic of your property.
3. **Landscaping:** Integrate storage systems into your landscaping design. Treat them as functional elements within the overall aesthetic, harmonizing nature and function. Your storage system is a part of a garden orchestra. Each element plays its role, contributing to a beautiful design.
4. **Accessibility:** Make sure your storage systems are easily accessible for maintenance. However, consider the visual aspects to maintain the property's appeal. It must be practical for maintenance but shouldn't distract from the property aesthetics.

The harmonious interplay of maintenance and integration is the key to a successful rainwater harvesting system. Regular upkeep ensures the efficient functioning of your system, while thoughtful integration enhances the visual appeal of your property. Like a skilled conductor guiding an orchestra, you direct the elements of your rainwater harvesting system to create a synchronization between sustainability and beauty.

As you conclude this exploration of storage systems in rainwater harvesting, it's evident that each component plays a distinct role, contributing to the overall harmony of sustainable water management. Whether it's the simplicity of gutters guiding the initial flow, the charm of barrels storing raindrops like precious notes, or the grandeur of tanks conducting a monumental tune, the choice of storage systems defines the rhythm of this water-harvesting composition.

Chapter 6: Safety and Filtration – Ensuring Clean Water for Every Use

Rainwater harvesting demands the assurance of clean and safe water for everyday use. This chapter focuses on the meticulous process of safeguarding harvested rainwater, addressing potential contaminants, exploring filtration techniques, and delving into post-storage treatments to guarantee water safety. From understanding the sources of contamination to implementing effective filtration methods and post-filtration treatments, this chapter will guide you through the comprehensive measures necessary for maintaining consistently clean water.

Rainwater harvesting demands the assurance of clean and safe water for everyday use.
https://pixabay.com/vectors/virus-boat-doctor-team-rescue-7341187/

Potential Contaminants in Harvested Rainwater

In the pursuit of sustainable water practices, harvested rainwater often emerges as a promising source. However, the apparent purity of rainwater can be deceiving. Before thinking about water safety, you need to understand the potential contaminants that infiltrate harvested rainwater. These contaminants, originating from various sources, span a spectrum that includes atmospheric pollutants, debris, and biological agents, forming a complex tapestry that requires careful consideration.

Atmospheric Pollutants: Unseen Impurities Descending from the Skies

On its descent, rainwater encounters a myriad of atmospheric pollutants that compromise its purity. These pollutants, though invisible to the naked eye, significantly impact the quality of harvested rainwater.

1. **Airborne Particulate Matter**: The seemingly innocent act of rain falling from the sky brings with it the accumulation of airborne particulate matter. Dust, pollen, and other microscopic particles suspended in the air settle on rooftops and are inevitably carried into harvested rainwater. The rooftops, once pristine, become catchment areas for these impurities, introducing a layer of complexity to the water harvesting process.

    - **Impact on Water Quality**: While these particles might seem inconsequential individually, their cumulative presence affects the taste and visual clarity of harvested rainwater. Moreover, they contribute to clogging in filtration systems if not addressed.
    - **Clogging Concerns**: Accumulation of airborne particles leads to clogging in downspouts and gutters, affecting the efficiency of rainwater collection.
    - **Visual Clarity**: The presence of particulate matter results in cloudy or turbid water, affecting the visual aesthetics of harvested rainwater.

2. **Chemical Pollutants**: The industrial landscape, a significant contributor to air pollution, casts a subtle but impactful shadow on harvested rainwater. Industrial emissions release a cocktail of chemicals into the atmosphere, some of which find their way into rainwater during its descent.

- **Sulfur Compounds**: Factories and industrial facilities emit sulfur compounds that dissolve in rainwater, potentially leading to the formation of acid rain. The presence of sulfur compounds in harvested rainwater alters its pH and introduces acidity.
- **Heavy Metals:** The insidious nature of heavy metals, such as lead, mercury, and cadmium, is exacerbated by their release into the air through industrial processes. Once airborne, these metals settle on surfaces, including rooftops, becoming unwelcome guests in harvested rainwater.
- **Impact on Water Quality:** Chemical pollutants introduce a range of undesirable characteristics to rainwater, from altered taste and color to potential health hazards associated with heavy metal ingestion.
- **Health Considerations**: Consumption of rainwater contaminated with heavy metals poses serious health risks, emphasizing the importance of effective filtration.

3. **Microorganisms:** The atmosphere, though vast and seemingly pure, is home to a myriad of microorganisms that hitch a ride with raindrops. Bacteria, viruses, and fungi present in the air find their way into harvested rainwater during its journey from the clouds to catchment surfaces.
    - **Impact on Water Quality:** While rainwater is generally considered free of harmful microorganisms, the potential introduction of these entities from the atmosphere emphasizes the need for thorough filtration and disinfection processes.
    - **Filtration Challenges**: Addressing microorganisms requires specialized filtration methods to ensure the removal of potential pathogens.
    - **Disinfection Considerations**: Microbial contamination highlights the importance of post-filtration disinfection to ensure the safety of harvested rainwater for various uses.

Debris and Environmental Factors: Ground-Level Challenges

Beyond atmospheric pollutants, ground-level factors contribute significantly to the potential contamination of harvested rainwater. These factors encompass a range of environmental challenges that merit

attention.
1. **Roof Material Contamination**: The type of material used for roofing plays a pivotal role in determining the quality of harvested rainwater. Roof materials leach substances that compromise water purity.
    - **Asbestos Contamination**: In older buildings, roofs made of asbestos release fibers into rainwater, posing health risks if consumed.
    - **Treated Wood:** Roofs constructed with treated wood introduce chemicals into harvested rainwater, adding another layer of complexity to safety considerations.
    - **Impact on Water Quality**: Roof material contamination underscores the importance of selecting roofing materials carefully, especially when aiming to harvest rainwater for potable use.
    - **Material Selection**: Choosing roofing materials with minimal leaching properties is crucial for maintaining water quality.
    - **Health Awareness:** Educating yourself about the potential risks associated with specific roofing materials promotes informed decision-making.
2. **Overhanging Trees**: The charm of overhanging trees comes with an unintended consequence for rainwater harvesting. Leaves, bird droppings, and other organic matter from trees become potential contaminants.
    - **Leaves:** Falling leaves introduce organic matter that decomposes in harvested rainwater, impacting its quality.
    - **Bird Droppings:** Bird droppings, while seemingly innocuous, harbor bacteria and contribute to microbial contamination.
    - **Impact on Water Quality:** The natural beauty of overhanging trees brings with it the responsibility of managing potential contaminants, requiring proactive measures to ensure water purity.
    - **Regular Pruning:** Trimming overhanging branches reduces the likelihood of leaves and debris entering the rainwater harvesting system.

- **Bird Deterrents**: Implementing measures to deter birds, such as installing bird spikes or nets, minimizes the introduction of bird-related contaminants.
3. **Animal Activity**: Birds, insects, and small animals find rooftops and catchment areas attractive, contributing to the potential contamination of harvested rainwater.
    - **Birds:** Besides droppings, birds bring in feathers, nesting materials, and even small prey, all of which impact water quality.
    - **Insects:** Insects, attracted to the moisture on rooftops, inadvertently become part of the harvested rainwater.
    - **Impact on Water Quality**: Managing the presence of animals on rooftops is essential to prevent their contribution to contaminants, emphasizing the need for protective measures.
    - **Mesh Barriers:** Installing mesh barriers or screens over gutters and downspouts prevents insects and larger debris from entering the system.
    - **Bird-Proofing**: Employing bird-proofing measures, such as installing deterrents or netting, reduces the likelihood of bird-related contamination.
4. **Runoff from Surfaces**: Adjacent surfaces, such as driveways or areas with polluted soil, contribute to runoff that finds its way into the rainwater harvesting system.
    - **Chemical Runoff:** Polluted soil or chemical-laden surfaces introduce substances into harvested rainwater.
    - **Sediment and Debris:** Runoff carries sediment and debris, adding to the challenges of water quality maintenance.
    - **Impact on Water Quality**: Managing runoff is a critical aspect of rainwater harvesting, requiring careful planning and measures to prevent the introduction of external contaminants.
    - **Permeable Surfaces:** Implementing permeable surfaces in the vicinity reduces runoff and minimizes the influx of external contaminants.

- **Rain Gardens:** Designing rain gardens or buffer zones absorbs and filters runoff before it reaches the rainwater harvesting system.

As you unravel the intricate web of potential contaminants in harvested rainwater, it becomes evident that ensuring water purity is a multi-faceted challenge. From atmospheric pollutants descending from the skies to ground-level factors contributing to runoff, each element requires careful consideration.

Harvesting rainwater holds incredible potential for sustainable water practices, but this potential can only be fully realized with a nuanced understanding of the challenges at hand. Addressing potential contaminants involves a combination of thoughtful infrastructure design, regular maintenance practices, and a commitment to ongoing water quality monitoring.

Filtration Techniques and Their Efficacy

Water safety involves navigating the intricate world of filtration techniques. Filtration is not a one-size-fits-all solution. It's a nuanced and multi-faceted process designed to address specific types of contaminants. From mesh filters that act as the first line of defense to sophisticated reverse osmosis systems capable of removing a broad spectrum of impurities, each filtration method plays a unique role in ensuring the purity of harvested rainwater.

Mesh Filters

1. **Mechanism:** Mesh filters, typically crafted from materials like stainless steel or nylon, operate on a simple yet effective principle. They physically block larger particles and debris from entering the water system.
    - **Mesh Material:** The choice of materials, such as stainless steel or nylon, ensures durability and resilience against environmental factors.
    - **Mesh Pore Size:** Variations in mesh pore size allow customization based on the size of particles to be filtered.
2. **Optimal Applications:** These filters find their sweet spot in scenarios where the focus is on removing larger particles. They serve as ideal pre-filters, stationed strategically to protect subsequent filtration stages from potential clogging.

- **Pre-Filtration**: Commonly employed as the first layer of defense, mesh filters prevent leaves, insects, and larger particulate matter from progressing further into the filtration system.
- **Rainwater Collection**: Mesh filters are integral to rainwater collection systems, ensuring that only clean water enters storage tanks.

3. **Efficacy:** While mesh filters exhibit high efficacy in trapping larger particles, their effectiveness wanes when it comes to dealing with dissolved pollutants or microorganisms.
   - **Limitations:** Mesh filters aren't the silver bullet for contaminants at the molecular or microbial level, necessitating additional filtration stages.
   - **Maintenance**: Regular cleaning is essential to prevent clogging and maintain optimal performance.
   - **Replacement Schedule:** Periodic replacement of mesh filters ensures continued effectiveness, especially in areas with high debris levels.

Sediment Filters: Navigating the Finer Particles

1. **Mechanism**: Sediment filters employ materials like sand or fabric to trap finer particles suspended in the water, offering a more nuanced approach to filtration.
   - **Material Variety:** The use of different materials provides flexibility to address specific particle sizes effectively.
   - **Depth Filtration:** Some sediment filters utilize depth filtration, enhancing their capacity to capture particles throughout the entire filter depth.
2. **Optimal Applications**: Effective in removing smaller sediment particles, silt, and fine debris, sediment filters carve a niche in scenarios where precision in particle removal is crucial.
   - **Fine Particle Removal:** Ideal for applications where the presence of fine sediment poses a challenge to water quality.
   - **Pre-RO Filtration:** Sediment filters often serve as precursors to more advanced filtration methods like reverse osmosis.

3. **Efficacy:** Sediment filters provide good filtration for sediments but aren't as adept at handling dissolved pollutants or microorganisms.
   - **Considerations**: You should be aware that sediment filters may need additional support to address contaminants beyond particulate matter.
   - **Maintenance:** Regular monitoring and replacement of sediment filters are necessary to prevent clogging and maintain efficiency.

Activated Carbon Filters
1. **Mechanism:** Activated carbon filters work by adsorbing organic compounds, chemicals, and some gases from the water, making them versatile tools in the filtration arsenal.
   - **Adsorption Power**: The porous structure of activated carbon enhances its ability to attract and trap impurities effectively.
   - **Microporous Structure:** Activated carbon possesses a microporous structure, providing a large surface area for adsorption.
2. **Optimal Applications:** Activated carbon filters shine in applications where the focus is on removing organic contaminants, chlorine, and specific chemicals that affect taste and odor.
   - **Organic Contaminant Removal**: Well-suited for scenarios where the water source is prone to organic impurities.
   - **Taste and Odor Improvement:** Activated carbon effectively enhances the taste and odor of harvested rainwater.
3. **Efficacy:** While activated carbon filters boast high efficacy for specific pollutants, they require regular replacement to maintain effectiveness.
   - **Replacement Schedule**: Follow manufacturer guidelines for replacement intervals for consistent performance.
   - **Cost Considerations:** The recurring cost of filter replacement should be factored into the overall system maintenance budget.
   - **Temperature Sensitivity:** The efficacy of activated carbon filters is influenced by water temperature, and you should consider this factor during system design.

UV Sterilization
1. **Mechanism:** UV sterilization harnesses ultraviolet light to disrupt the DNA of microorganisms, rendering them unable to reproduce and guaranteeing a microbial-free water supply.
    - **Microbial Disruption**: The targeted use of UV light effectively neutralizes bacteria, viruses, and other microorganisms.
    - **UV-C Wavelength:** UV sterilization typically utilizes UV-C light, which is particularly effective in microbial inactivation.
2. **Optimal Applications**: UV sterilization shines in scenarios where microbial disinfection is the primary concern, offering a technology-driven solution for water safety.
    - **Microbial Threat Mitigation**: Ideal for applications where the risk of microbial contamination is high.
    - **Post-Filtration Disinfection:** UV sterilization is often employed as a post-filtration step to ensure the microbiological safety of harvested rainwater.
3. **Efficacy:** Highly effective for microbial disinfection and UV sterilization. However, it falls short when it comes to removing particulate matter or chemical contaminants.
    - **Additional Filtration:** Combining UV sterilization with other filtration methods addresses the comprehensive removal of impurities.
    - **Energy Consumption**: Consider the energy requirements associated with UV sterilization systems.
    - **Installation Considerations:** Proper installation and regular maintenance are crucial for the sustained efficacy of UV sterilization systems.

Reverse Osmosis
1. **Mechanism**: Reverse osmosis employs a semipermeable membrane to remove ions, molecules, and larger particles from the water, offering a comprehensive solution to diverse contaminants.
    - **Membrane Technology**: The semipermeable membrane acts as a molecular sieve, allowing only pure water molecules to pass through.

- **Pressure Differential**: Reverse osmosis relies on a pressure differential to drive the separation of water from impurities.
2. **Optimal Applications:** Reverse osmosis shines in applications where a wide range of contaminants, including minerals, chemicals, and microorganisms, need to be effectively removed.
    - **Diverse Contaminant Removal:** Suitable for scenarios where water purity is paramount, addressing mineral content, chemicals, and microbial threats.
    - **Residential Use:** Commonly employed in residential settings for producing purified drinking water.
3. **Efficacy**: Highly efficient in removing diverse contaminants, reverse osmosis systems have lower water production rates and involve wastewater generation.
    - **Water Production Rate**: Users should be mindful of the potential impact on water availability, especially in areas with limited rainfall.
    - **Wastewater Considerations**: Reverse osmosis systems generate wastewater, and proper disposal or reuse strategies should be in place.
    - **Mineral Removal:** While effective in removing minerals, you may need to consider remineralization methods for the produced water.

The effectiveness of filtration techniques lies in their application. From the robust defense of mesh filters against larger particles to the precision of reverse osmosis in tackling a spectrum of impurities, each method contributes to water safety.

Harvesting rainwater for consumption or various domestic purposes demands a tailored approach. A well-designed filtration system, incorporating the strengths of different techniques, ensures the removal of specific contaminants and the overall purity and safety of the harvested rainwater.

Disinfection and Regular Maintenance

Filtration is the vanguard in harvesting clean rainwater, but the quest for purity doesn't end there. Post-storage treatments and regular maintenance emerge as unsung heroes, ensuring the sustained quality of harvested rainwater. In this section, you'll discover various disinfection methods, such as chlorination and ozonation. Additionally, you'll see the intricacies of regular maintenance practices that form the backbone of a resilient rainwater harvesting system.

Chlorination

1. **Mechanism:** Chlorination, a time-tested method, involves the introduction of chlorine or chlorine-based compounds into water. This chemical disinfection process is designed to annihilate microorganisms and safeguard water quality.
   - **Chlorine Types:** Chlorine is applied in various forms, including chlorine gas, liquid sodium hypochlorite, or solid calcium hypochlorite.
   - **Microbial Neutralization:** Chlorine disrupts the cellular structures of bacteria and viruses, rendering them inactive and preventing waterborne diseases.
2. **Optimal Applications:** Chlorination finds its sweet spot in scenarios where continuous disinfection of stored water is paramount.
   - **Tank Disinfection:** Applied to water stored in tanks, chlorination ensures a consistent level of microbial control.
   - **Public Water Systems:** Widely used in municipal water treatment, chlorination is a staple for ensuring potable water.
3. **Considerations:** While chlorination is a potent disinfection method, careful dosing is crucial to prevent over-chlorination.
   - **Dosing Control:** Precise control of chlorine dosage is essential to avoid health risks associated with excess chlorine in drinking water.
   - **Chemical Contaminant Limitations:** Chlorination may not effectively remove certain chemical contaminants, necessitating additional filtration measures.
   - **Residual Chlorine Management:** Managing residual chlorine levels is crucial to ensure water safety and prevent undesirable taste or odor.

Ozonation

1. **Mechanism:** Ozonation introduces ozone, a powerful oxidizing agent, into water. Ozone's oxidative prowess disinfects water by

neutralizing contaminants and microorganisms.

- **Ozone Generation:** Ozone is generated on-site using specialized ozone generators, ensuring freshness and efficacy.
- **Pathogen Inactivation:** Ozone effectively inactivates bacteria, viruses, and some organic pollutants, contributing to comprehensive water safety.

2. **Optimal Applications:** Ozonation shines in applications where a broader spectrum of contaminants, including bacteria and viruses, must be addressed.

- **Microbial Disinfection:** Ozone serves as a robust defense against pathogenic microorganisms, ensuring waterborne disease prevention.
- **Organic Pollutant Removal:** Ozone's ability to break down organic pollutants enhances its efficacy in improving water quality.

3. **Considerations:** Implementing ozone systems requires careful calibration and attention to specific considerations to ensure optimal performance.

- **Calibration Precision:** Ozone systems demand precise calibration to achieve the desired disinfection efficacy without compromising water safety.
- **Residual Ozone Management:** Managing residual ozone levels is crucial, as excessive ozone in the water is harmful and impacts the water's taste.
- **Complexity of Installation:** Ozonation systems, while effective, are complex to install and may require professional assistance.

Regular Maintenance Practices

Ensuring the long-term safety of harvested rainwater involves the diligent integration of regular maintenance practices. These practices form a proactive shield against potential threats to water quality, guaranteeing a consistent supply of clean and safe rainwater.

Tank Inspection
- **Regularity:** Tanks should be routinely inspected for signs of wear, corrosion, or damage that could compromise water quality.
- **Seal Integrity:** Make sure that tank seals and joints are intact, preventing the infiltration of external contaminants.
- **Coating Integrity:** Check the integrity of any coatings on the tank interior, addressing any degradation promptly.

Cleaning Gutters and Screens
- **Frequency:** Regularly clean gutters and mesh filters to prevent the buildup of debris that compromises water quality.
- **Mesh Integrity:** Closely monitor the mesh integrity of filters, repairing or replacing damaged sections promptly.
- **Efficient Water Flow:** Unobstructed gutters and screens maintain efficient water flow, reducing the risk of contamination.

Filter Replacement
- **Adherence to Schedule:** Replace filters as recommended by the manufacturer to maintain their efficacy.
- **Filter Type Consideration:** Consider the specific type of filter and its lifespan, adjusting the replacement schedule accordingly.
- **Documentation:** Keep a record of filter replacements to facilitate a proactive and organized maintenance approach.

Flushing the System
- **Periodicity:** Periodically flush the system to remove stagnant water and potential contaminants.
- **System Efficiency:** Flushing enhances system efficiency by preventing the buildup of sediments and microbial growth.
- **Water Quality Assurance:** Regular flushing contributes to consistent water quality, especially in systems with infrequent use.

The essence of water purity lies in diligence and proactive care. Chlorination and ozonation stand as stalwart guardians, neutralizing threats at the chemical and microbial levels. Regular maintenance practices ensure the long-term integrity of the rainwater harvesting system.

The synergy of these approaches echoes the principles of sustainability and environmental stewardship. The journey doesn't end with the collection of rainwater. It extends into the treatment and maintenance, guaranteeing that every drop harvested remains a testament to the commitment to clean and sustainable water practices.

By addressing these aspects, you'll harness the full potential of rainwater harvesting, not only as a sustainable water source but also as a source of water purity. The commitment to water safety is not just a technical endeavor. It is a holistic approach that considers environmental factors, technology, and proactive maintenance.

Chapter 7: Beyond the Basics - Advanced Systems and Techniques

Water harvesting has evolved far beyond the rudimentary practices of the past. In this chapter, you'll learn about cutting-edge technologies, innovative materials, and integrated systems that define the forefront of rainwater harvesting. These advanced methods optimize water collection and storage, seamlessly integrating with other sustainable practices and offering a holistic approach to water management in various climates and terrains.

Innovative Materials and Designs

Rainwater harvesting, once a simple practice reliant on basic roofing materials, has entered a new era of innovation. Recent advancements in materials and designs are reshaping how people collect and utilize rainwater. In this section, you'll explore three groundbreaking innovations that are transforming rainwater harvesting.

Aerogels

Aerogels are revolutionary materials with a porous and lightweight structure that enhances water collection efficiency to new heights.
https://commons.wikimedia.org/wiki/File:Aerogel_hand.jpg

Traditionally, the effectiveness of rainwater harvesting depended on the design of the collection surface. Enter aerogels, a revolutionary material with a porous and lightweight structure that enhances water collection efficiency to new heights.

- **Increased Surface Area for Enhanced Collection**: Aerogels, with their intricate structure, provide an expanded surface area, capturing more water droplets from the air than conventional materials.

- **Versatility for Retrofitting**: The lightweight nature of aerogels makes them an ideal choice for integration into existing roofing materials, allowing for easy retrofitting of structures.

- **Rapid Condensation for Maximum Yield**: The porous structure of aerogels facilitates rapid condensation, ensuring that even the smallest droplets are gathered efficiently, resulting in a substantial boost in overall water yield.

Smart Surfaces

Imagine surfaces that respond intelligently to environmental conditions, optimizing the entire rainwater harvesting process. Smart surfaces equipped with sensors and actuators make this vision a reality.

- **Real-Time Environmental Monitoring:** Smart surfaces are embedded with sensors that detect changes in temperature, humidity, and precipitation in real-time, allowing for adaptive responses.
- **Optimal Water Flow and Filtration:** Actuators on smart surfaces adjust properties like inclination and permeability, facilitating optimal water flow during heavy rainfall and enhanced filtration during lighter precipitation.
- **Remote Monitoring for Efficiency:** Integration with the Internet of Things (IoT) allows users to monitor the efficiency of their rainwater harvesting systems remotely, ensuring proactive maintenance and optimal water quality.

The marriage of aerogels and smart surfaces represents a leap forward in rainwater harvesting technology. It combines the efficiency of advanced materials and the adaptability of intelligent design, promising greater yields and sustainability.

Hydrophilic Coatings

One of the challenges in rainwater harvesting, especially in arid regions, is the low humidity that limits water collection. Hydrophilic coatings present a game-changing solution by giving surfaces the ability to attract and retain water molecules from the air.

- **Versatility for Various Materials:** Hydrophilic coatings can be applied to a variety of materials, including traditional roofing materials, offering a scalable solution for enhancing efficiency.
- **Promoting Water Adhesion:** The coatings create an environment that encourages water molecules to adhere, significantly increasing water capture even in low-humidity conditions.
- **Cost-Effective and Scalable:** The adaptability of hydrophilic coatings makes them a cost-effective and scalable solution for enhancing the efficiency of existing rainwater harvesting systems.

In addition to increasing water capture, hydrophilic coatings contribute to the prevention of water runoff. By promoting water adhesion, these coatings minimize wastage and maximize the potential for collection.

Architectural Marvels

In the world of rainwater harvesting, form is now meeting function through cutting-edge architectural designs. The integration of rainwater collection systems into buildings has given rise to structures that are both practical and aesthetically pleasing.

- **Self-Draining Roofs and Facades:** Innovative designs incorporate self-draining roofs that efficiently channel rainwater toward collection points, eliminating stagnant water and potential leaks.
- **Visual Harmony with Greenery**: Rooftops adorned with lush greenery contribute to a visually harmonious and eco-friendly environment where plants assist in water absorption and purification.
- **Interactive Building Elements:** Some architectural marvels go beyond passive rainwater collection, incorporating interactive elements that engage inhabitants in the water harvesting process.

Now that you have an idea of the technological marvels in the modern world, here's a closer look at each of them.

1. **Self-Draining Roofs and Facades:** Imagine a building with a roof that protects from the elements and actively contributes to water collection. Self-draining roofs are designed to channel rainwater efficiently toward collection points, eliminating stagnant water and potential leaks.
   - **Efficient Water Channeling:** These roofs are equipped with a slope and drainage system that ensures rainwater is directed towards collection points, preventing water from pooling and causing damage.
   - **Elimination of Stagnant Water:** By efficiently draining rainwater, self-draining roofs eliminate the risk of stagnant water, reducing the potential for leaks and structural damage.
   - **Enhanced Durability:** The design enhances water collection and contributes to the longevity of the roofing system, as pooling water is a common cause of deterioration.
2. **Visual Harmony with the Environment**: Architectural designs incorporate rainwater collection elements seamlessly into the visual appeal of the structure. Rooftops adorned with lush greenery, where plants assist in water absorption and purification, contribute to a visually harmonious and eco-friendly environment.
   - **Aesthetic Integration:** Rainwater collection elements are seamlessly integrated into the overall design of the building,

enhancing its aesthetic appeal and contributing to a visually pleasing environment.
- **Green Rooftops for Biodiversity:** Rooftop gardens aid in rainwater absorption and create habitats for biodiversity, promoting ecological balance in urban environments.
- **Dual Functionality:** The integration of aesthetics with the rainwater collection provides dual functionality, making the building visually appealing and environmentally conscious.

3. **Interactive Building Elements:** Some architectural marvels go beyond passive rainwater collection. They incorporate interactive elements, such as transparent sections that showcase the flow of rainwater or kinetic features that respond to the volume of collected water. These designs serve a practical purpose and engage inhabitants in the water harvesting process.
   - **Transparency for Education:** Transparent sections in building elements allow inhabitants to observe the flow of rainwater, promoting awareness and education about the importance of water conservation.
   - **Kinetic Features for Engagement:** Kinetic elements, such as water features activated by collected rainwater, provide an engaging and interactive experience for inhabitants, fostering a sense of connection with the water harvesting process.
   - **Educational and Recreational Value:** Interactive building elements contribute to the efficiency of rainwater harvesting and add educational and recreational value to the building, enhancing its overall significance in the community.

Advanced materials and designs are ushering in a new era for rainwater harvesting. Aerogels and smart surfaces offer unprecedented efficiency and adaptability. Hydrophilic coatings tackle the challenge of low-humidity environments. Architectural marvels redefine how you perceive the integration of water collection into your built environment. As you embrace these innovations, you move toward a more sustainable and water-secure future.

Automation in Rainwater Harvesting

In the ever-evolving landscape of water management, automation has emerged as a beacon of efficiency, transforming rainwater harvesting into a smart and sustainable practice. In this section, you'll explore the three pillars of automation in rainwater harvesting. These advancements streamline the collection and storage process while paving the way for a more conscientious and resource-efficient approach to water utilization.

IoT Sensors

The integration of IoT into rainwater harvesting marks a paradigm shift in how people approach water resource management. IoT sensors, strategically placed on rooftops and within storage systems, serve as the eyes and ears of the entire system. They continuously gather real-time data on crucial parameters such as rainfall, water levels, and overall system health.

- **Real-Time Data for Informed Decision-Making:** These sensors empower users with a wealth of information at their fingertips. You'll access live data on the current rainfall intensity or the precise volume of water stored in your harvesting system from the comfort of your smartphone or computer. This real-time insight allows for informed decision-making, enabling you to optimize your harvesting systems based on current conditions.

- **Proactive Maintenance for System Longevity:** One of the significant advantages of IoT sensors is their ability to facilitate proactive maintenance. By monitoring the health of the system in real time, you identify potential issues before they escalate. Whether it's a clogged filter, a malfunctioning pump, or a structural concern, early detection allows for timely intervention, preserving the efficiency and longevity of the rainwater harvesting infrastructure.

- **Accessibility and Remote Monitoring:** The accessibility of real-time data is not limited by geographical constraints. You can remotely monitor your rainwater harvesting systems, making it especially beneficial for installations in remote or hard-to-reach locations. This capability enhances the overall efficiency of system management by enabling quick responses to changing conditions, regardless of physical proximity.

Automated Filtration Systems

Rainwater, while a valuable resource, is not immune to impurities. Automated filtration systems equipped with advanced technologies are revolutionizing the way you ensure the purity of harvested water.

- **Self-Cleaning Mechanisms for Uninterrupted Performance:** Traditional filtration systems often require manual intervention for cleaning and maintenance. Automated filtration systems, on the other hand, feature self-cleaning mechanisms that keep the filters free from contaminants. It guarantees a continuous supply of clean water and minimizes the need for frequent and labor-intensive upkeep.
- **Adapting to Varying Water Qualities:** Water quality varies based on factors such as climate, seasonal changes, and environmental influences. Automated filtration systems are designed to adapt dynamically to these variations. Whether the water source is experiencing a sudden influx of debris during heavy rainfall or a change in composition during dry spells, these systems adjust their filtration processes to maintain optimal performance.
- **Integration with Water Quality Sensors:** The automation of filtration systems can be further enhanced through integration with water quality sensors. These sensors detect specific impurities or contaminants in the harvested water. The filtration system then adjusts its processes in response to real-time data, ensuring that the water meets the desired quality standards. This level of precision in water purification enhances the overall reliability and safety of the harvested water.

Smart Distribution Networks

Automation doesn't end with collection and storage. It extends its transformative touch to the distribution phase. Smart distribution networks leverage automation to regulate the flow of water based on demand and availability.

- **Regulating Flow with Smart Pumps and Valves**: In traditional rainwater harvesting setups, water distribution often relies on manual adjustments or fixed schedules. Smart pumps and valves, powered by automation, bring a new level of precision to this process. These intelligent components regulate the flow of water based on real-time demand, ensuring a constant and reliable supply.
- **Minimizing Energy Consumption for Sustainability:** Energy efficiency is a cornerstone of sustainable water management. Smart distribution networks excel in this aspect by minimizing energy consumption. Pumps and valves operate precisely when

needed, avoiding unnecessary energy expenditure. It contributes to environmental sustainability and translates into cost savings for you.

- **Predictive Analytics for Optimal Resource Allocation:** The integration of predictive analytics into smart distribution networks adds another layer of sophistication. By analyzing historical usage patterns and considering environmental factors, these systems predict future demand with remarkable accuracy. This predictive capability allows for proactive adjustments in water distribution, optimizing resource allocation and ensuring a consistent supply even during periods of high demand.

The marriage of IoT sensors, automated filtration systems, and smart distribution networks forms a trinity that propels rainwater harvesting into a realm of holistic sustainability. This synergy optimizes the efficiency of water collection and ensures the purity of the harvested water and its judicious distribution.

As you embrace the era of automation in rainwater harvesting, you pave the way for a future where water is managed with precision, sustainability, and a profound understanding of its vital role in your life. The journey towards water resilience has never been more promising.

Integrated Sustainable Systems

In the ever-evolving landscape of sustainable water management, the integration of various systems yields powerful and synergistic results. This section will shine a light on three interconnected approaches that exemplify the symbiosis between rainwater harvesting and other sustainable practices.

Greywater Recycling

Rainwater harvesting, when coupled with greywater recycling, establishes a holistic water management system that maximizes every precious drop. Greywater, originating from domestic activities like bathing and laundry, is a valuable resource often underutilized. The integration of greywater with rainwater harvesting creates a harmonious relationship where both water sources complement each other.

- **Tapping into Greywater's Potential:** While greywater is not suitable for drinking, its potential for non-potable uses is immense. Properly treated, greywater can be seamlessly combined with harvested rainwater for applications like irrigation.

This dual approach significantly reduces reliance on external water sources for non-drinking purposes, a pivotal step towards sustainable water utilization.
- **Treatment and Purification Strategies**: Greywater treatment involves the removal of impurities and contaminants to meet the required quality standards for its intended use. Technologies such as filtration and biological treatment systems are employed to purify greywater. When merged with harvested rainwater, this treated blend becomes a versatile and eco-friendly water source for maintaining gardens, lawns, and other non-potable water needs.
- **Reducing Environmental Impact:** Beyond its immediate benefits, integrating greywater recycling with rainwater harvesting can potentially reduce the environmental impact associated with water consumption. By minimizing reliance on conventional water sources, this combined system contributes to the conservation of freshwater ecosystems and mitigates the strain on municipal water supplies.
- **Accessibility and Remote Monitoring:** The accessibility to real-time data is not limited by geographical constraints. You can remotely monitor your rainwater harvesting systems. This capability enhances the overall efficiency of system management by enabling quick responses to changing conditions, regardless of physical proximity.

Permaculture Designs

Permaculture, deeply rooted in principles of sustainability and harmonious coexistence with the environment, offers a natural companion to rainwater harvesting. Integrating these practices allows for the creation of self-sustaining ecosystems that promote soil health, biodiversity, and overall environmental well-being.
- **The Role of Rainwater in Permaculture:** Rainwater, as a fundamental component of permaculture designs, is harnessed to nurture the landscape. Various techniques, such as swales and contour bunds, are employed to capture and direct rainwater to where it is needed most. It guarantees efficient water utilization and prevents soil erosion while aiding in the cultivation of diverse plant species.

- **Regenerative Agriculture Practices**: The integration of rainwater harvesting and permaculture extends to regenerative agriculture practices. You'll foster healthier soil conditions by capturing rainwater and implementing permaculture techniques. This, in turn, enhances crop resilience, reduces the need for synthetic fertilizers, and contributes to the restoration of degraded land.
- **Biodiversity Promotion:** Permaculture emphasizes the importance of biodiversity in agricultural and natural systems. A more resilient and diverse ecosystem can be created by incorporating rainwater harvesting into permaculture designs. The stored rainwater provides a lifeline during dry spells, fostering the survival of various plant and animal species.
- **Community Engagement and Education:** The integration of permaculture and rainwater harvesting is not only about cultivating sustainable ecosystems but also about community engagement and education. By sharing knowledge and practices, communities can collectively work towards a more sustainable and regenerative way of living.

Aquaponics and Rainwater

The marriage of aquaponics and rainwater harvesting presents an innovative approach to sustainable agriculture. In aquaponics, fish and plants form a symbiotic relationship where fish waste provides essential nutrients for plants. The plants, in turn, purify the water for the fish. When seamlessly integrated with rainwater harvesting, this system becomes a closed-loop, resource-efficient powerhouse.

- **Rainwater as the Lifeblood of Aquaponics**: Rainwater, collected and stored, serves as the nutrient-rich lifeblood of aquaponics systems. You'll reduce your dependence on external water sources by utilizing rainwater in aquaponics.
- **Resource Efficiency and Closed-Loop Systems**: The integration of aquaponics and rainwater harvesting embodies the essence of resource efficiency. Fish waste, a natural byproduct of aquaponics, becomes a fertilizer for plants. As the plants absorb these nutrients, they contribute to the purification of the water, creating a closed-loop system that minimizes waste and maximizes efficiency.
- **A Blueprint for Urban Agriculture**: The compact nature of aquaponics makes it particularly suitable for urban agriculture.

By incorporating rainwater harvesting, you create sustainable and self-sufficient growing systems. This approach reduces the environmental footprint associated with conventional agriculture and provides a blueprint for cultivating fresh produce in limited urban spaces.

- **Educational Opportunities and Food Security**: The integration of aquaponics and rainwater harvesting also offers educational opportunities and contributes to food security. By promoting this sustainable farming method, you learn about the interdependence of ecosystems and gain valuable skills in urban agriculture, fostering a sense of food autonomy.

Smart Landscaping

Among the many facets of integrated sustainable systems, smart landscaping emerges as a pivotal player, seamlessly blending aesthetics with purpose. By incorporating rainwater harvesting into landscaping designs, you create outdoor spaces that captivate the eye and contribute to environmental sustainability.

- **Rain Gardens and Sustainable Landscapes:** Rain gardens, strategically designed to capture and manage stormwater runoff, exemplify the union of landscaping and rainwater harvesting. These gardens are engineered to absorb rainwater, preventing soil erosion and minimizing the flow of contaminants into water bodies. By integrating rain gardens into landscaping plans, you'll transform your outdoor spaces into dynamic elements of a sustainable water management system.
- **Edible Landscapes and Urban Agriculture**: The concept of edible landscapes takes landscaping a step further, integrating ornamental plants with edible ones. By combining rainwater harvesting with edible landscaping, you'll cultivate fruits, vegetables, and herbs using collected rainwater. This approach provides a sustainable source of fresh produce and reduces the carbon footprint associated with transporting food from distant locations.
- **Biodiverse and Water-Efficient Plant Selection:** Smart landscaping extends to the careful selection of plants that thrive in specific climates with minimal water requirements. By choosing native and drought-resistant plants, you'll contribute to water efficiency and biodiversity. This intentional selection aligns

with the principles of permaculture, promoting a harmonious relationship between human-managed spaces and the natural environment.

As you explore the integration of rainwater harvesting with greywater recycling, permaculture designs, aquaponics, and smart landscaping, a common thread emerges. It's the pursuit of harmony with the environment. These integrated sustainable systems optimize water utilization and foster regenerative practices that contribute to the planet's well-being.

Toward a Sustainable Future

The journey toward a sustainable future involves embracing diverse practices that work in tandem to conserve resources and nurture ecosystems. The integration of rainwater harvesting with greywater recycling, permaculture designs, and aquaponics exemplifies this harmonious approach. By adopting these integrated systems, individuals, communities, and agricultural practitioners can become stewards of a more sustainable and resilient world.

The future of water management lies in synergy. Innovative materials and designs enhance the efficiency of collection structures. Automation streamlines the entire process, and integration with other sustainable practices creates holistic solutions. The adaptability of these advancements makes them applicable in various climates and terrains, empowering you to harness the power of rainwater most efficiently and sustainably.

Chapter 8: Nature's Bounty – Uses for Your Harvest

Rainwater harvesting opens the door to a wealth of possibilities, turning every droplet into a valuable resource. In this chapter, you'll explore the myriad applications of harvested rainwater, underscoring its versatility and the potential to redefine how you meet your water needs. From everyday household uses to agricultural benefits and the possibilities of potable water, you'll uncover the vast potential nature's bounty holds.

Distinguishing Potable and Non-Potable Uses

Before exploring the diverse applications of harvested rainwater, it's crucial to distinguish between potable and non-potable uses. Potable water is suitable for drinking, while non-potable water is used for other purposes, such as irrigation or cleaning. The water quality required for these applications varies, with potable water demanding the highest standards to guarantee human health.

To transform harvested rainwater into a safe drinking source, stringent quality parameters must be met. It involves thorough filtration, disinfection, and monitoring to eliminate contaminants and pathogens. Compliance with regulatory standards ensures that the water is potable and meets the highest safety requirements.

Household Applications

In sustainable living, harvested rainwater emerges as a versatile and invaluable resource, revolutionizing everyday household applications. From laundry to cleaning and gardening, rainwater's soft and chemical-free composition transforms mundane chores into eco-friendly and economically sustainable practices. Here's how rainwater can enhance the fabric of your daily life.

Laundry

Laundry day takes on a whole new dimension with the introduction of harvested rainwater.
https://pixabay.com/vectors/washhouse-laundry-house-room-294621/

Laundry day takes on a whole new dimension with the introduction of harvested rainwater. Unlike hard water from conventional sources, rainwater is naturally soft, devoid of harsh minerals that compromise the effectiveness of detergents and take a toll on fabrics. This inherent softness creates a gentle, nurturing environment for clothing, ensuring a softer touch and prolonging the life of each garment.

- **Enhancing Detergent Efficiency:** The soft nature of rainwater enhances the efficiency of detergents, allowing them to lather more effectively and penetrate fabric fibers with ease. Even in areas with hard tap water, where detergents struggle to reach their full potential, rainwater provides a solution that maximizes cleaning power while minimizing the use of chemical additives.

- **Prolonging Clothing Longevity:** Hard water, laden with minerals like calcium and magnesium, contributes to the wear and tear of clothing over time. The abrasive effects of these minerals on fabric fibers lead to fading, reduced softness, and a shorter lifespan for garments. By opting for rainwater in laundry, you'll invest in the longevity of your clothing, reducing the frequency of replacements and contributing to a more sustainable wardrobe.

- **Reducing Environmental Impact:** Beyond the benefits to fabrics, using rainwater for laundry contributes to environmental sustainability. Traditional water sources often require extensive treatment processes and transport, consuming energy and contributing to carbon emissions. In contrast, rainwater harvesting for laundry significantly reduces the environmental impact associated with water use, aligning with eco-conscious living practices.

Cleaning

Household cleaning takes on a new dimension when nature's purifier, rainwater, becomes the cleaning agent of choice. Its soft and chemical-free composition makes it ideal for a variety of cleaning purposes, from surfaces and windows to delicate items that require a tender touch.

- **Ideal for Surfaces and Windows:** The softness of rainwater makes it particularly effective for cleaning surfaces and windows. Without the minerals present in hard water that leave streaks and residue, rainwater leads to a crystal-clear finish. Whether it's wiping down countertops, glass tables, or mirrors, rainwater leaves surfaces spotless, all while being gentle on the materials being cleaned.

- **Preserving Delicate Items:** Certain delicate items, such as intricate glassware, fragile decorations, or heirloom pieces, benefit from the gentle touch of rainwater. The absence of harsh chemicals ensures that delicate surfaces remain unharmed during the cleaning process. It preserves the integrity of these items and reflects a commitment to sustainable cleaning practices.

- **Reducing Reliance on Tap Water:** Harnessing rainwater for cleaning needs reduces reliance on tap water, contributing to both environmental and economic sustainability. The energy-intensive processes involved in treating and distributing tap water are minimized, resulting in a reduced carbon footprint. This shift towards rainwater for cleaning aligns with the broader movement toward responsible water use.

Gardening

Perhaps one of the most rewarding applications of harvested rainwater is in the garden. Rainwater, free from chlorine and other additives commonly found in tap water, nurtures plants with the pure essence of hydration. The controlled pH levels in rainwater make it a perfect match for various plant species, promoting healthy growth and vibrant blooms.

- **Chlorine-Free Hydration**: Many municipal water supplies are treated with chlorine to eliminate bacteria and pathogens. While this is essential for human consumption, plants do not share the same affinity for chlorine. Rainwater, being free from this chemical additive, provides plants with a chlorine-free source of hydration, promoting optimal growth and development.

- **Controlled pH Levels for Plant Health**: Rainwater typically has a slightly acidic pH, which is beneficial for certain plants that thrive in acidic soil conditions. This controlled pH level ensures that rainwater complements the specific needs of a variety of plant species, fostering an environment where they flourish. This natural compatibility makes rainwater ideal for watering gardens and potted plants.

- **Promoting Water Efficiency in Gardening**: Water efficiency in gardening is a key consideration for sustainable practices. Rainwater harvesting directly addresses this concern by providing a local and on-site water source for plants. By utilizing rainwater for gardening, you'll reduce the demand for municipal water supplies, contributing to water conservation efforts and promoting a more resilient and eco-friendly landscape.

As you navigate the waters of sustainable living, the incorporation of harvested rainwater into household applications reveals a gentle yet transformative touch. From the softer embrace it offers to fabrics to its role as nature's purifier in cleaning and the growth it conducts in the garden, rainwater emerges as a precious resource that goes beyond mere functionality.

Agricultural and Landscaping Benefits

In the vast tapestry of sustainable water management, the agricultural and landscaping sectors stand as canvas and garden alike, awaiting the transformative touch of harvested rainwater. This precious resource, harvested and stored during moments of abundance, is a lifeline for crops and a nurturing elixir for landscapes. This section scours the agricultural realm, where rainwater becomes a sustainable alternative for farming, and explores how landscaping transforms into natural masterpieces under its gentle care.

Farming

In the agricultural expanse, water is the lifeline that sustains crops, ensuring their growth, health, and productivity. Traditional sources, however, often come with challenges of fluctuating water availability, reliance on distant reservoirs, and the need for energy-intensive irrigation systems. This is where rainwater harvesting steps in as a sustainable alternative, offering a locally sourced and environmentally friendly solution.

- **Safeguarding Crops from Drought Stress:** The unpredictable nature of weather patterns, including periods of drought, poses a significant threat to agricultural productivity. Rainwater harvesting provides a buffer against these challenges by enabling farmers to store and utilize rainwater during dry spells. This stored bounty becomes a lifeline for crops, offering a consistent water supply that safeguards them from the stress of water scarcity.

- **Enhancing Soil Health with Pure Rainwater**: Beyond just providing hydration, rainwater contributes to the overall health of agricultural lands. Unlike traditional water sources treated with chlorine and other additives, rainwater is pure and free from chemical interventions. This purity extends to the soil, enhancing its health and fertility. The absence of chemical residues ensures that the soil becomes a thriving ecosystem where beneficial microorganisms flourish, supporting the vitality of crops.

- **Promoting Sustainable Agriculture Practices:** Rainwater harvesting aligns seamlessly with the principles of sustainable agriculture. By relying on a locally sourced and naturally replenished water supply, you'll reduce your dependence on external water sources. It minimizes the environmental impact associated with long-distance water transport and promotes a more self-sufficient and resilient agricultural system.

Landscaping

Landscaping, whether in residential gardens or expansive public spaces, transforms into a canvas of natural masterpieces when nourished by harvested rainwater. The purity and softness of rainwater offer a gentle touch that promotes the health and vibrancy of plants. Free from the harsh minerals and additives found in conventional water sources, rainwater becomes a nurturing elixir for the green elements of your surroundings.

- **Reducing the Need for Chemical Interventions:** Conventional water sources often contain minerals that, over time, accumulate in the soil and affect plant health. Harsh elements like calcium and magnesium present in hard water necessitate chemical interventions to counter their impact. Rainwater, with its soft and mineral-free composition, eliminates the need for such interventions, allowing plants to thrive naturally.
- **Efficient Irrigation for Thriving Landscapes**: The controlled distribution of rainwater through efficient irrigation systems is a key element in crafting natural masterpieces in landscaping. Rainwater harvesting systems, equipped with intelligent irrigation mechanisms, ensure that landscapes receive just the right amount of water. This precision promotes water efficiency, preventing overwatering and minimizing runoff. It contributes to an eco-friendly and sustainable landscaping approach.
- **Enhancing Biodiversity and Ecological Balance**: The soft touch of rainwater extends beyond individual plants to the broader ecosystem within landscapes. By reducing the reliance on conventional water sources, which are treated with chemicals for purification, rainwater contributes to the preservation of biodiversity. Beneficial insects, microorganisms, and other components of the ecological balance within the landscape thrive in an environment free from the adverse effects of chemical-laden water.

The synergy between rainwater harvesting in agriculture and landscaping creates integrated solutions that enhance water efficiency on multiple fronts. The same harvested rainwater that nourishes crops is directed to landscaping elements, fostering a harmonious approach to water use. This integration aligns with the principles of permaculture, where diverse elements coexist and contribute to a self-sustaining ecosystem.

The adoption of rainwater harvesting in these sectors is not merely a practical choice but a commitment to a future where water is valued. It's where ecosystems flourish, and landscapes become vibrant expressions of ecological harmony. By embracing rainwater as a cornerstone of agricultural and landscaping practices, you'll cultivate a sustainable legacy that respects the delicate balance of nature and ensures a resilient future for generations to come.

Potable Possibilities

The delicate alchemy of transforming harvested rainwater into potable elixir is a journey guided by stringent quality assurance measures. Filtration systems, ultraviolet treatments, and unwavering compliance with regulatory standards become the sentinels. These ensure that the pure essence of rainwater emerges as a thirst-quencher and as a paragon of safety.

Quality Assurance for Drinking Water

The metamorphosis of rainwater into potable water demands precision and dedication to quality. From the serene descent of raindrops to the moment it cascades into a drinking glass, every step is full of responsibility. The commitment to quality assurance transforms rain's embrace into a source of life that not only hydrates but nourishes without compromise.

- **Filtration Systems**: Harvested rainwater, while pristine in its origin, carries microorganisms and chemicals that pose potential risks to human health. Robust filtration systems are the first line of defense, serving as guardians against contaminants. Mesh filters, adept at trapping larger particles, and carbon filters, capable of removing impurities and odors, form a formidable alliance. These filters work in harmony to ensure that the water undergoes a transformative cleansing, emerging free from visible and invisible intruders.
- **Ultraviolet Treatments:** Beyond the physical aspects of filtration lies the illuminating touch of ultraviolet (UV) treatments. UV disinfection becomes a crucial step in the purification process. It targets microorganisms that persist despite the initial filtration. The power of UV light disrupts the DNA of bacteria, viruses,

and other pathogens, rendering them incapable of causing harm. Once touched by the sun's rays, this final purification ensures that the water emerges as a microbiologically safe liquid.
- **Regulatory Standards**: In the vast expanse of potable possibilities, adherence to regulatory standards becomes the North Star, guiding the entire process. Local health regulations outline the benchmarks for water quality. They set the stage for a process where safety is non-negotiable. Understanding and incorporating these standards into the purification process guarantees the final product quenches thirst without compromising health.

Microbial and Chemical Safety Measures

The purity that makes rainwater a source of wonder also introduces challenges. Harvested rainwater carries microorganisms and chemicals that, if left unchecked, compromise its safety for human consumption. Recognizing and addressing these challenges becomes imperative in crafting potable possibilities from rain's embrace.

- **Robust Filtration Systems:** The journey toward potability begins with robust filtration systems that serve as the first line of defense. Mesh filters, with their intricate weave, capture larger particles and debris, preventing them from tainting the water. Carbon filters, with their porous structure, adsorb impurities, odors, and some chemicals, further enhancing the clarity and purity of the water. Together, these filtration systems form a formidable barrier against contaminants.
- **UV Disinfection:** The microbial safety measures reach their pinnacle with UV disinfection—a process that harnesses the power of light to neutralize microorganisms. The short-wavelength UV-C light damages the DNA of bacteria, viruses, and other pathogens, rendering them unable to reproduce or cause infection. This layer of protection ensures that the final product is not just visually clear but microbiologically safe.
- **Regular Testing:** The journey from raindrop to drinking glass is an ongoing commitment to safety. Regular testing for microbial and chemical parameters becomes a continuous assurance of the water's quality. Rigorous testing protocols, including checks for bacteria, viruses, and chemical compositions, guarantee that the potable possibilities of harvested rainwater remain steadfast in their commitment to health and safety.

As you savor the potable possibilities of harvested rainwater, let it be a reminder of the delicate balance between nature's bounty and your responsibility to safeguard health. In each sip, you taste not just the freshness of rain but the culmination of a journey. From the cloud-kissed skies to the vigilant filtration systems, from the dance of UV light to the commitment to regulatory standards, it's a journey where the possibilities of rainwater extend beyond nourishment. This liquid gold becomes a celebration of purity, a testament to human ingenuity, and a harmonious dance with the essence of life itself.

Environmental and Economic Benefits

In the delicate balance between nature's bounty and human needs, the harvesting of rainwater emerges as a transformative practice, ushering in a wave of environmental and economic benefits. At the heart of this practice lies the profound impact of reducing strain on municipal water supplies, mitigating stormwater runoff, and unveiling a water-wise approach that extends financial savings to the water-wise wallet.

- **Conserving Treated Water for Essential Uses**: One of the primary benefits of harvested rainwater lies in its potential to alleviate the strain on municipal water supplies. By meeting non-potable needs with naturally sourced water, you'll contribute significantly to the conservation of treated water for essential uses. This prudent conservation approach ensures that the limited and precious resource of treated water is reserved for purposes that demand the highest quality.

- **Easing the Burden on Water Treatment Facilities**: As rainwater becomes a readily available source for activities like gardening, cleaning, and irrigation, the burden on water treatment facilities eases. These facilities are designed to purify water to meet rigorous drinking standards. By diverting non-potable demands to harvested rainwater, you play a proactive role in preserving the integrity of treated water. It optimizes the efficiency of treatment processes and extends the lifespan of water infrastructure.

- **Reducing Energy Demands for Water Transportation**: The journey of water from treatment facilities to homes involves significant energy consumption, especially when transported over long distances. Harvested rainwater, sourced locally, disrupts this energy-intensive cycle. Utilizing water where it falls reduces the need for extensive transportation, contributing to a more sustainable and energy-efficient water supply system.

Proactive Management of Excess Water

Beyond its role in reducing strain on municipal water supplies, rainwater harvesting plays a crucial role in mitigating stormwater runoff. Stormwater runoff, often a culprit in urban flooding, results from rainfall that exceeds the absorption capacity of soil and surfaces. Harvesting rainwater at its source transforms homeowners and communities into proactive managers of excess water. You'll prevent soil erosion and minimize the flow of pollutants into rivers and streams.

- **Preserving Soil Health and Water Purity:** As rainwater is captured and directed for various uses, it permeates into the soil, replenishing aquifers and preserving soil health. It mitigates the risk of soil erosion and prevents the rapid flow of rainwater over impermeable surfaces, reducing the chances of water pollution. By becoming a steward of rain's descent, you'll embrace a holistic approach to water management that safeguards both the environment and its inhabitants.

- **Shrinking Water Bills through Smart Choices**: The economic benefits of harvesting rainwater extend beyond the environmental realm, offering tangible savings to households. As reliance on municipal water for non-potable needs diminishes, so do water bills. The initial investment in a rainwater harvesting system becomes a wise and enduring economic choice, creating a sustainable and cost-effective water supply.

- **Long-Term Economic Gain:** While there is an initial investment in installing rainwater harvesting systems, the long-term economic gain is substantial. The reduction in water bills, coupled with the potential for local government incentives or rebates for sustainable practices, transforms rainwater harvesting into a financially savvy choice. As you witness your water-wise wallet grow, the economic viability of rainwater harvesting becomes increasingly apparent.

Maximum Yield

When you get into sustainable water management, harvesting maximum yield from rainwater presents itself as a transformative and empowering practice. At its core, this journey encompasses the implementation of efficient rainwater harvesting systems, the optimization of storage capacities, and the embrace of water-wise practices. The union of environmental consciousness with practical application fosters individual empowerment and cultivates a community of stewards dedicated to making the most of every precious drop.

- **Designing for Maximum Capture**: At the heart of harvesting maximum yield is the thoughtful design and implementation of efficient rainwater harvesting systems. The journey begins with capturing rainwater at its source, be it on rooftops, surfaces, or catchment areas. Thoughtfully designed systems, equipped with advanced technologies and materials, ensure every raindrop is harnessed with precision.
- **Strategic Storage Solutions:** Optimizing harvested yield involves strategic storage solutions that align with the natural rhythms of rainfall. Robust storage capacities, whether in above-ground tanks, cisterns, or below-ground reservoirs, become the reservoirs of abundance. By maximizing storage, you store surplus rainwater for periods of scarcity, guaranteeing a consistent and reliable water supply throughout the year.
- **Smart Distribution Networks:** Equally important is the establishment of smart distribution networks within the storage system. Intelligent pumps, valves, and distribution mechanisms ensure that the stored rainwater is distributed efficiently, addressing the specific needs of different areas, whether for irrigation, gardening, or non-potable household uses. This strategic distribution optimizes the utility of harvested rainwater, maximizing its impact across various facets of daily life.
- **Landscape Design and Irrigation Efficiency:** Harvesting maximum yield extends beyond the technicalities of systems and storage. It embraces water-wise practices that cultivate conscious consumption. Thoughtful landscape design, incorporating native and drought-resistant plants, minimizes water demands. Efficient irrigation systems, such as drip irrigation or rain garden techniques, make it so that every drop is utilized properly, promoting a harmonious balance between nature and human needs.
- **Indoor Conservation Measures**: Water-wise practices also find their place indoors, where conscious consumption becomes a daily commitment. Installing low-flow fixtures, fixing leaks

promptly, and embracing water-efficient appliances contribute to the overall goal of maximizing the utility of harvested rainwater. These measures amplify the impact of rainwater harvesting on reducing dependence on conventional water sources.

- **Workshops and Seminars**: Obtaining maximum yield is not a solitary endeavor. It thrives on educational initiatives and community engagement. Workshops and seminars become platforms for sharing insights into the benefits and applications of harvested rainwater. These provide you with the knowledge and tools needed to become an active participant in the journey toward sustainability.
- **Community Outreach Programs:** The ripple effect of change is amplified through community outreach programs. These initiatives foster a sense of community responsibility, encouraging individuals to become catalysts for positive change within their neighborhoods. By collectively embracing the principles of rainwater harvesting, communities transform into stewards of their water resources, nurturing a shared commitment to environmental sustainability.

Water-wise practices in outdoor landscaping and indoor consumption become the threads that weave conscious living. Yet, the journey is incomplete without the communal spirit fostered by educational initiatives and community engagement. As workshops and community outreach programs empower individuals with the knowledge to become stewards of their water resources, the collective impact becomes a force for positive change.

In concluding this exploration of nature's bounty and the uses for your harvest, the overarching theme is one of harmony and sustainability. From everyday household applications to agricultural benefits and the possibilities of potable water, harvested rainwater is a versatile and valuable resource. As you nurture nature's gift drop by drop, you take a step closer to a more sustainable, resilient, and water-wise future.

Chapter 9: Potable Rainwater: Making Your Harvest Drinkable

In the pursuit of self-sufficiency and sustainable living, transforming harvested rainwater into safe, potable water stands as a pinnacle of achievement. This chapter explores the intricate processes and necessary precautions required to make your rainwater harvest usable and drinkable. From understanding local water quality to employing advanced filtration methods and disinfection techniques, the journey toward potable rainwater is an exploration of both science and practicality.

In the pursuit of self-sufficiency and sustainable living, transforming harvested rainwater into safe, potable water stands as a pinnacle of achievement.
https://www.pexels.com/photo/crop-person-filling-bottle-with-water-from-drinking-fountain-7245245/

Understanding Local Water Quality

Every region, with its unique blend of environmental influences, industrial activities, and human settlements, has distinct challenges and characteristics that shape its water quality. This section explores the crucial significance of understanding these dynamics, emphasizing the awareness of potential contaminants, both natural and anthropogenic, as the foundational step toward ensuring the safety of the final drinking product.

Local Water Quality Dynamics

Nature's influence, industrial activities, and human settlements collectively shape the dynamic tapestry of local water quality. Understanding these intricate dynamics lays the groundwork for anticipating potential contaminants and tailoring purification strategies to the unique characteristics of each region.

- **Environmental Influences**: Nature, in all its diversity, plays a pivotal role in shaping the quality of local water sources. Environmental factors such as soil composition, topography, and vegetation contribute to the mineral content and overall characteristics of rainwater. For instance, water flowing through rocky terrain might carry higher mineral concentrations, impacting taste and safety. Understanding these natural influences provides a baseline for anticipating potential challenges in the purification process.

- **Industrial Activities:** Human activities, especially industrial processes, introduce a spectrum of substances into the local water supply. Runoff from industrial areas carries pollutants such as heavy metals, chemicals, and toxins. Awareness of nearby industrial activities is crucial in identifying potential contaminants that could seep into rainwater. This insight directs the selection of appropriate filtration and purification methods to address specific industrial pollutants.

- **Human Settlements**: Urban and rural settlements leave their imprints on local water quality. Urban areas introduce pollutants like pesticides, herbicides, and contaminants from vehicle emissions. In contrast, rural areas see agricultural runoff carrying fertilizers and pesticides into water sources. Understanding the footprint of human settlements enables a tailored approach to water purification, addressing the unique challenges posed by each environment.

Testing Protocols and Frequency

Testing is the guardian of the promise of safe drinking water. This section explores the essential parameters, from microbial content to chemical composition, and emphasizes the importance of a vigilant testing regime. Adhering to local regulations, adapting to seasonal variations, and monitoring changes in the surrounding environment are integral components of this ongoing commitment. A comprehensive testing regime should cover a spectrum of parameters critical to water quality:

- **Microbial Content:** Testing for bacteria, viruses, and other microorganisms assesses the microbial safety of the water. Coliform bacteria, for example, serve as indicators of fecal contamination and potential pathogenic risks.
- **Chemical Composition:** Analyzing the chemical makeup detects substances such as heavy metals, pesticides, and industrial pollutants. This step is vital in addressing both natural and anthropogenic contaminants.
- **Overall Water Quality:** Parameters like pH, turbidity, and dissolved oxygen contribute to the overall quality and usability of the water. Maintaining these within acceptable ranges ensures a safe and pleasant drinking experience.

Frequency of Testing

The frequency of testing is a dynamic aspect that adapts to local conditions and regulations:

- **Local Regulations:** Adherence to local regulations is paramount. Some regions have specific guidelines dictating the testing frequency for different parameters. Understanding and complying with these regulations establish a legal framework for ensuring water safety.
- **Seasonal Variations:** Seasonal changes influence water quality too. Increased agricultural activities during planting seasons or industrial processes in certain weather conditions elevate contamination risks. Adjusting the testing frequency to align with these variations ensures a proactive approach to potential challenges.

- **Changes in Surrounding Environment:** Environmental shifts, such as nearby construction, changes in land use, or new industrial developments, impact water quality. Regular testing, especially during periods of environmental change, is an early warning system, allowing for prompt adjustments to purification methods.

A vigilant testing regime is a guardian that upholds the promise of safe drinking water. It is not a one-time affair but an ongoing commitment to monitor and adapt to the dynamic nature of local water quality. Regular testing serves as a proactive measure, allowing for timely adjustments in purification methods and ensuring the sustained potability of harvested rainwater.

Advanced Filtration Methods

Efficient filtration is the linchpin in the ambitious journey to transform rainwater into a potable resource. As you delve into the intricacies of advanced filtration methods, you'll encounter a world where the understanding of the size of microns becomes paramount. It's time to explore the science behind microns and their role in bacteria removal. You'll also uncover the pivotal contributions of activated carbon filters and reverse osmosis systems in ensuring a comprehensive and purified drinking experience.

Micron Sizes and Bacteria Removal

Understanding the size of microns is pivotal in designing filtration systems that act as the first line of defense against bacteria. Microfiltration and ultrafiltration systems, with their distinct pore sizes, lay the groundwork for thorough bacteria removal, ensuring the journey toward potable rainwater starts with precision.

- **The Microscopic World:** At the heart of advanced filtration is the microscopic realm of microorganisms, particularly bacteria. These tiny living beings, with sizes ranging from 0.2 to 5 microns, pose a significant challenge in the quest for potable rainwater. To effectively remove these threats, filtration systems must be strategically designed to capture particles within this size range.
- **Microfiltration:** Microfiltration systems, characterized by their relatively larger pore sizes compared to other advanced methods, serve as the initial defense against microorganisms. These filters typically have pores ranging from 0.1 to 10 microns, making

them adept at trapping larger particles like bacteria. However, their efficacy varies, and additional filtration methods are often needed to ensure thorough removal.
- **Ultrafiltration**: Taking filtration precision to the next level, ultrafiltration systems boast smaller pore sizes, typically ranging from 0.002 to 0.1 microns. That enables them to capture even the smallest bacteria, providing a more comprehensive solution for microbial removal. Ultrafiltration, with its ability to target particles at the sub-micron level, lays the groundwork for achieving the stringent standards required for safe drinking water.

Activated Carbon Filters

Beyond microbial challenges, rainwater carries an array of chemical contaminants. Activated carbon filters step into the spotlight with their porous prowess, excelling at absorbing chemicals like chlorine, pesticides, and organic compounds. This section explores the magic of adsorption, the versatility of activated carbon in addressing various contaminants, and how this dual-action purification elevates the quality of harvested rainwater.

- **The Porous Powerhouse:** Activated carbon filters emerge as the unsung heroes in the battle against chemical impurities. Their porous structure, created through a process that activates carbon with steam or chemicals, provides an expansive surface area for adsorption.
- **Adsorption Magic:** Activated carbon's adsorption capacity is a game-changer in the purification process. As water passes through the filter, chemical contaminants adhere to the carbon surface, effectively removing them from the water. This dual-action purification, addressing both microbial and chemical impurities, elevates the quality of harvested rainwater to meet the high standards required for safe consumption.
- **Versatility in Application**: Activated carbon filters prove versatile in addressing a wide spectrum of chemical contaminants, including:
    - **a. Chlorine**: Commonly used in water treatment but undesirable in drinking water due to taste and potential health concerns.
    - **b. Pesticides and Herbicides**: Agricultural runoff can introduce these chemicals into rainwater, posing risks to human health.

c. **Organic Compounds:** Various pollutants from industrial activities find their way into rainwater, necessitating effective removal for safety.

Reverse Osmosis Systems

In the pursuit of a comprehensive purification approach, reverse osmosis (RO) systems emerge as key players. Operating at the molecular level, RO utilizes a semi-permeable membrane to filter out impurities, from bacteria to dissolved salts and minerals. Here's how RO works:

1. **Semi-Permeable Membrane**: The heart of the RO system, the semi-permeable membrane, allows water molecules to pass through while blocking larger contaminants.
2. **Pressure Application:** Applying pressure to the water forces it through the membrane, separating impurities and pollutants.
3. **Reject Water Disposal**: The concentrated contaminants are then flushed away as rejected water, leaving behind purified water ready for consumption.

Understanding micron sizes is the compass guiding you through the microscopic world of bacteria removal, while activated carbon filters showcase their prowess in adsorbing chemical contaminants, ensuring a dual-action purification. Reverse osmosis systems provide a comprehensive purification approach that transcends bacterial threats to address molecular-level impurities. The combined efforts of these advanced filtration methods elevate harvested rainwater to a standard of potable purity, marking a triumph in the quest for sustainable and safe drinking water.

Disinfection Techniques

In the relentless pursuit of transforming harvested rainwater into potable water, the spotlight shifts to disinfection techniques. It's a critical phase where microbial safety takes center stage. From harnessing the power of light to embracing age-old practices, these techniques stand as guardians, ensuring the journey from raindrop to drinking glass is free from microbial threats.

UV Purification

In the radiant realm of modern disinfection, UV purification is a powerful and effective method. As you unveil the science behind harnessing the power of light, specifically UV-C, you'll witness a process that damages the DNA of microorganisms, preventing them from reproducing and causing infections. Integrated into the water distribution network, UV systems provide continuous disinfection without altering the water's taste or introducing additional chemicals.

- **The Radiant Solution**: UV purification is a testament to the power of light in neutralizing microbial threats. Specifically, UV-C light, with its wavelength between 200 and 280 nanometers, becomes the weapon of choice. As rainwater flows through UV systems integrated into the water distribution network, a silent yet powerful process unfolds.
- **DNA Damage**: UV-C light, when targeted at microorganisms, wreaks havoc at the molecular level. It damages the DNA of bacteria, viruses, and other pathogens, rendering them incapable of reproduction. This disruption in the life cycle ensures that even if microorganisms survive exposure to UV light, they cannot proliferate or cause infections. The result is a water supply that is continuously disinfected without altering its taste or introducing additional chemicals.
- **Integration into Water Distribution**: The seamless integration of UV systems into the water distribution network is a hallmark of their effectiveness. As rainwater makes its way through pipes and conduits, UV-C lights stand guard, providing a continuous and automated disinfection process. This integration ensures the microbial safety of the water at the point of consumption and minimizes the need for manual intervention, making UV purification a reliable and efficient safeguard.

Chlorination

Journeying into the annals of water treatment history, you'll encounter the enduring legacy of chlorination. A method that has withstood the test of time, chlorination involves the addition of chlorine to water for disinfection. Residual chlorine monitoring is a key focus, highlighting chlorination's time-tested efficacy.

- **The Enduring Legacy of Chlorine**: Chlorination is a time-tested approach with a legacy spanning over a century. The principle is simple yet effective. Chlorine, in various forms such as chlorine gas, sodium hypochlorite, or calcium hypochlorite, is a potent agent against a broad spectrum of microorganisms.

- **Broad-Spectrum Disinfection:** Chlorine's efficacy lies in its ability to eliminate not only bacteria but also viruses, algae, and other pathogens. It disrupts the life cycle by attacking cellular structures and enzymes, rendering them unable to function. The result is a comprehensive disinfection that safeguards against a wide range of potential threats in harvested rainwater.
- **Dosage Control:** While chlorination is a powerful disinfection method, the key lies in precise dosage. Adding too many compromises the taste and safety of the water, while adding too little fails to achieve effective disinfection. Achieving this balance requires careful monitoring and control of chlorine levels throughout the water distribution network.
- **Residual Chlorine Monitoring:** Maintaining residual chlorine levels becomes a key aspect of the chlorination process. Residual chlorine, the amount of chlorine that remains in the water after disinfection, is an indicator of ongoing microbial protection. Regular monitoring ensures that the water continues to meet safety standards without compromising taste. It's a delicate equilibrium that underscores the importance of chlorination's time-tested efficacy.

Boiling

As you embrace tradition in modern times, boiling takes center stage as a simple yet highly effective method of sterilizing water. Whether it's the straightforward mechanism of pathogen eradication through heat or the altitude considerations that shape boiling practices, the simplicity of boiling is the epitome of effectiveness.

- **Embracing Tradition in Modern Times:** In situations where advanced technologies aren't readily available, boiling is the best practice. Boiling, an age-old practice rooted in tradition, remains a simple yet highly effective method of sterilizing water. As rainwater reaches its boiling point of 100 degrees Celsius (212 degrees Fahrenheit), most pathogens are eradicated.
- **Pathogen Eradication through Heat:** The mechanism is straightforward. The application of heat through boiling disrupts the structural integrity of microorganisms. While boiling doesn't remove chemical contaminants, it provides a practical and accessible means of ensuring microbial safety. This simplicity becomes especially valuable in situations where access to

sophisticated water treatment infrastructure is limited.
- **Altitude Considerations:** In regions with higher altitudes, where water boils at lower temperatures due to reduced atmospheric pressure, the recommended boiling time is extended to ensure complete pathogen eradication. Boiling for at least one minute (or three minutes at higher altitudes) becomes the golden rule, reaffirming the simplicity and effectiveness of this age-old practice.

From the radiant elegance of UV purification to the time-tested legacy of chlorination and the simplicity of boiling, each technique stands as a sentinel, guaranteeing that harvested rainwater reaches its final destination free from microbial threats. In this process of disinfection, tradition meets innovation, and simplicity intertwines with sophistication, creating a harmonious journey from the heavens to the human thirst.

Importance of Regular Testing and Maintenance

In the journey from raindrop to drinking glass, where advanced filtration and disinfection methods stand as guardians, regular testing and maintenance are paramount. Post-treatment testing and system maintenance ensure the sustained safety and quality of harvested rainwater. From the delicate balance of residual chlorine levels to the vigilant checks on microbial content, this iterative process is the lifeline that safeguards water quality over time.

Post-Treatment Testing

After the rainwater undergoes advanced filtration and disinfection, post-treatment testing illuminates the unseen threats that linger despite the formidable defenses of UV purification and chlorination. From the delicate balance of residual chlorine levels to the vigilant checks on microbial content and overall water quality, post-treatment testing ensures the continued safety and quality of harvested rainwater.

- **Residual Chlorine Levels:** When using the chlorination method, achieving the balance of residual chlorine levels is critical. Residual chlorine is your tell against microbial resurgence. Too little, and the water becomes vulnerable to contamination. On the other hand, with too much chlorine, the taste and safety of the water are compromised.

- **Microbial Content:** Microbial content testing delves into the microscopic universe, where unseen pathogens persist. Even the most advanced filtration methods leave behind traces of microorganisms. Post-treatment testing ensures that these invisible threats are exposed and neutralized, reinforcing the microbial safety of rainwater.
- **Overall Water Quality:** Beyond individual components, overall water quality testing provides a comprehensive evaluation. Parameters such as pH, turbidity, and dissolved oxygen contribute to the holistic understanding of water quality. Regular assessments guarantee that the water remains pleasant to the taste and free from undesirable characteristics.
- **Iterative Refinement:** The iterative nature of post-treatment testing is the key to sustaining water quality over time. It's not a one-time validation but a continual refinement process. As environmental conditions change, seasonal variations occur, and the water distribution network evolves, regular testing adapts to these dynamics, ensuring that the safety standards set for rainwater are consistently met.

System Maintenance

Filters, UV lamps, and the entire rainwater treatment system require vigilant attention to maintain their effectiveness. This section explores how regular maintenance ensures that the gatekeepers of purity continue to uphold their role in removing impurities and microorganisms. It also sheds light on the importance of periodic checks and replacements for UV lamps, those that safeguard against microbial threats.

- **Filters:** Filters, the gatekeepers of purity in the rainwater treatment process, require vigilant attention. Over time, they accumulate particles and contaminants, diminishing their effectiveness. Regular cleaning or replacement ensures that the filtration system continues to uphold its role in removing impurities and microorganisms.
- **UV Lamps:** In UV purification, the effectiveness of UV lamps is pivotal. These lamps emit powerful UV-C light that damages the DNA of microorganisms. Periodic checks and, if necessary, replacements guarantee that the UV purification system remains a formidable barrier against microbial threats.

- **System Vigilance**: Neglecting system maintenance means you're leaving the gates unguarded. As filters become clogged and UV lamps dim, the entire rainwater harvesting and treatment system becomes susceptible to a decline in effectiveness. A compromised system jeopardizes water quality and poses risks to human health.
- **The Proactive Approach:** System maintenance is not a reactive response to issues. It's a proactive approach to sustaining the integrity of the entire rainwater treatment infrastructure. Regular checks, scheduled replacements, and keeping an eye on the overall system health become the proactive measures that prevent potential problems before they compromise the safety of the drinking water.

Post-treatment testing and system maintenance embody a continual commitment to purity. It's a pledge to safeguard the journey from raindrops to drinking glass against unseen threats and system wear. In this commitment, tradition meets innovation, and simplicity intertwines with sophistication. It creates a harmonious balance that ensures sustained safety and quality of harvested rainwater for generations to come.

Understanding local water quality sets the stage for a targeted approach, while advanced filtration methods tailor the purification process for potability. Disinfection techniques, whether through UV purification, chlorination, or boiling, lead to safety. Regular testing and maintenance should be the cornerstone for sustaining the promise of safe drinking water.

Chapter 10: A Sustainable Future – Techniques in Modern-Day Conservation

In the face of escalating global challenges such as water scarcity and climate change, rainwater harvesting represents hope for a sustainable future. This final chapter explores the contemporary significance of rainwater harvesting and the latest innovations in sustainable water use. You'll discover how it integrates into broader conservation efforts. As you navigate the intricate landscape of modern-day conservation, this chapter aims to inspire you to see your rainwater harvesting endeavors as integral contributions to a global movement towards environmental stewardship.

Understanding Global Challenges

Water scarcity, once a concern limited to specific regions, has now evolved into a critical global issue. The surge in urbanization, coupled with population growth and inefficient water management practices, has placed an unprecedented strain on this planet's water resources. In the face of this growing crisis, rainwater harvesting offers a decentralized and sustainable approach to augmenting water supplies.

Water scarcity, once a concern limited to specific regions, has now evolved into a critical global issue.
Genetics4good, GFDL <http://www.gnu.org/copyleft/fdl.html>, via Wikimedia Commons: https://commons.wikimedia.org/wiki/File:Water_stress_2019_WRI.png

The catalyst for this global imperative is none other than climate change. This phenomenon has brought about significant alterations in precipitation patterns, an increase in the frequency of extreme weather events, and an exacerbation of water scarcity in various regions. To address these challenges, rainwater harvesting stands as a mitigation strategy and a resilient response to the unpredictable shifts in climate patterns.

The Escalating Crisis of Water Scarcity

The intertwining factors of rapid urbanization and population growth have drastically increased the water demand. In many regions, traditional water sources are unable to meet this surging demand, leading to water scarcity that extends beyond geographic boundaries. The urgency of the situation is magnified by inefficient water management practices that further deplete available water resources.

Unlike centralized water supply systems, which often struggle to cope with increasing demand, rainwater harvesting provides a decentralized solution. By capturing and utilizing rainwater at the local level, communities can reduce their reliance on overburdened water infrastructure and tap into a sustainable source that replenishes with each rainfall.

Climate Change

The impacts of climate change on water resources are profound and multifaceted. Altered precipitation patterns lead to irregular water availability, making it essential for communities to adapt their water management strategies. Extreme weather events, such as droughts and floods, further intensify the challenges of water scarcity, emphasizing the need for immediate and proactive measures.

By capturing rainwater, communities can build resilience against the uncertainties associated with shifting climate patterns. The decentralized nature of rainwater harvesting aligns seamlessly with the call for adaptive measures in the face of climate change.

Innovations in Sustainable Water Use

As the world grapples with the pressing challenge of water scarcity, innovative solutions are emerging to revolutionize the way people use and manage water. One such avenue of progress lies in sustainable construction and agricultural practices. Green building designs and advanced irrigation methods are reshaping the global approach to water conservation.

Green Building Designs

In the quest for sustainable water use, conservation efforts are extending beyond the conventional. Modern architects are embracing a paradigm shift by seamlessly integrating nature into building designs. Green building designs go beyond aesthetics by transforming structures into sustainable ecosystems. Rooftop gardens, permeable surfaces, and self-draining structures are all elements of a holistic approach to water conservation.

Rooftop gardens, for instance, serve a dual purpose. They enhance rainwater collection by providing a natural surface for water accumulation and contribute to urban biodiversity. These green oases create habitats for plants and insects, fostering a healthier urban ecosystem. Additionally, permeable surfaces allow rainwater to infiltrate the ground, replenishing aquifers and reducing surface runoff that can lead to flooding.

The integration of self-draining structures is another innovation in green building designs. These structures are designed to efficiently manage rainwater, directing it away from buildings and into collection systems. By doing so, they mitigate the urban heat island effect, contributing to a cooler and more sustainable urban environment.

Advanced Irrigation Methods

Agriculture, a sector that consumes a significant portion of the world's water supply, is undergoing a transformative revolution in irrigation methods. Precision agriculture, fueled by technology, is at the forefront of this change. The key objective is to deliver water precisely where and when it is needed, optimizing usage and minimizing waste.

The integration of rainwater into advanced irrigation systems further enhances their efficiency. By capturing and utilizing rainwater, farmers are reducing their reliance on traditional water sources, mitigating the impact on local ecosystems. This approach fosters a more sustainable model of food production.

Precision agriculture utilizes sensors, data analytics, and automated systems to monitor and manage crop conditions. This data-driven approach allows farmers to make informed irrigation decisions, optimizing water use for maximum crop yield. By embracing precision agriculture and incorporating rainwater into these systems, humanity is moving towards a more sustainable and water-efficient future for global agriculture.

Synergistic Environmental Benefits Through Integration

The true potential of rainwater harvesting lies in its ability to synergize with other conservation practices. This section explores the transformative impact of integrating rainwater harvesting with techniques such as greywater recycling, permaculture designs, and sustainable landscaping. It's time for you to understand how creating a holistic ecosystem extends beyond immediate water supply concerns.

Greywater Recycling

Rainwater harvesting, when seamlessly integrated with greywater recycling, forms a powerful duo in sustainable water management. Greywater, derived from daily activities such as laundry and bathing, complements rainwater by providing an additional source for non-potable uses. By combining these two sources, communities can significantly reduce their dependence on traditional water supplies, easing the burden on strained water resources.

Greywater recycling systems capture, treat, and reuse water that would otherwise go to waste. When this recycled water is synchronized with rainwater harvesting, it creates a cyclical and efficient water management system. This synergy promotes a more sustainable lifestyle, highlighting the interconnectedness of various water sources.

Permaculture Designs

Integrating rainwater harvesting with permaculture designs takes the concept of sustainability to another level. Permaculture principles guide the creation of self-sustaining environments that mimic natural ecosystems. By harmonizing rainwater harvesting with permaculture, you create landscapes that foster biodiversity, enrich soil health, and promote regenerative agriculture.

Permaculture emphasizes working with nature instead of against it. Through careful design, rainwater is directed to nourish plants, support food forests, and create microclimates that enhance overall ecosystem resilience. This approach contributes to the creation of vibrant and sustainable living environments.

Ecosystem Restoration

The impact of rainwater harvesting extends far beyond meeting immediate water needs. It becomes a catalyst for ecosystem restoration, playing a crucial role in preserving natural habitats. By replenishing groundwater levels and supporting the health of rivers and lakes, rainwater harvesting contributes to the overall well-being of ecosystems.

When integrated with other conservation practices, rainwater harvesting becomes a force for positive change. It nurtures the resilience of entire ecosystems, ensuring the health of flora and fauna that depend on sustainable water sources. This interconnected approach is a reminder that your efforts are not just about securing water for today but about creating a legacy of environmental stewardship for generations to come.

A Global Movement Toward Sustainability

In the quest for sustainability, many countries are recognizing the profound impact of their actions. Embracing rainwater harvesting as a broader conservation narrative transforms people from passive observers to active participants in a global movement toward sustainability.

Individual Actions, Global Impact

Rainwater harvesting, when embraced on an individual level, goes beyond personal water security. It becomes a cornerstone of a global movement towards sustainability. The cumulative effect of countless individuals adopting rainwater harvesting practices has a profound impact.

It creates a network of interconnected efforts that transcend geographic boundaries. It influences the health of entire ecosystems and contributes to the larger narrative of environmental responsibility.

Individuals who embrace rainwater harvesting recognize that their actions are part of a broader ecosystem. By capturing and utilizing rainwater, you contribute to the conservation of traditional water sources, alleviating the strain on local water supplies. It secures water for personal use and safeguards the delicate balance of ecosystems that depend on sustainable water sources.

Community Engagement and Advocacy

The power of rainwater harvesting extends beyond its immediate utility. It catalyzes community engagement and advocacy. Individuals who have experienced the benefits of rainwater harvesting often become passionate advocates for sustainable water practices. Sharing success stories, promoting awareness, and collaborating on larger conservation initiatives create a ripple effect that amplifies the impact of rainwater harvesting.

Communities that come together to embrace rainwater harvesting initiate a positive feedback loop. As awareness spreads, more individuals are inspired to adopt these practices, creating a groundswell of support for sustainable water management. This community-driven approach strengthens local resilience and contributes to a broader cultural shift toward sustainability.

Educational Initiatives: Shaping Future Stewards

The journey toward a sustainable future involves education and empowerment. Rainwater harvesting provides an excellent opportunity to integrate sustainability into educational curricula and community outreach programs. Incorporating this practice into the learning experience will cultivate a new generation of environmental stewards.

Educational initiatives centered on rainwater harvesting go beyond theory. They provide practical knowledge that empowers individuals to make a tangible difference. As students and community members learn about the environmental impact of their choices, they become active contributors to a sustainable future. These future leaders will carry the torch forward, ensuring that the ethos of sustainability becomes an integral part of the collective consciousness.

The global movement towards sustainability is not an abstract concept but a collective effort built on individual actions. Rainwater harvesting, when embraced by individuals and communities, becomes a powerful force in this movement. From securing personal water needs to influencing the health of ecosystems and advocating for broader environmental conservation, the ripple effect of rainwater harvesting is shaping a more sustainable world. As you educate, engage, and advocate, you cultivate a legacy of environmental stewardship for generations to come.

Conclusion: A Call to Action

As you conclude this exploration into rainwater harvesting and modern-day conservation, it's essential to recognize the transformative potential within your grasp. Rainwater harvesting is more than just a technique. It's a philosophy that recognizes the interconnectedness of human actions with the health of the planet.

7 Ways Harvesting Rainwater is Beneficial for the Future

From ensuring water security to fostering ecosystem resilience, rainwater harvesting is a versatile and essential practice. Here are some ways in which integrating sustainability and conservation techniques amplifies the benefits of rainwater harvesting, creating a harmonious approach to resource management.

1. **Water Security and Independence**

    The primary and most immediate benefit of rainwater harvesting is the assurance of water security. As populations grow and traditional water sources get drained, capturing rainwater provides a decentralized and reliable water supply. Rooftop harvesting systems, for example, allow individuals and communities to collect rainwater for various uses, from domestic needs to agricultural irrigation.

    Water independence is particularly crucial in regions with unreliable infrastructure or vulnerable to drought. By harvesting rainwater, you can mitigate the impact of water shortages, ensuring a continuous and reliable water source, even in arid climates.

2. **Mitigating the Urban Heat Island Effect**

   Urban areas often experience higher temperatures than their rural counterparts, creating what is known as the urban heat island effect. Rainwater harvesting, especially when integrated into green building designs, contributes to mitigating this heat island effect.

   Green roofs and permeable surfaces, commonly associated with rainwater harvesting practices, provide shade, reduce surface temperatures, and improve overall urban microclimates. By lessening the heat island effect, rainwater harvesting contributes to creating more comfortable and sustainable urban environments.

3. **Biodiversity Enhancement through Sustainable Landscaping**

   Rainwater harvesting extends beyond collecting water. It involves a holistic approach to landscaping that enhances biodiversity. Sustainable landscaping practices that integrate rainwater harvesting create environments that support a variety of plants and animal life.

   Utilizing rainwater for landscaping reduces the reliance on traditional irrigation methods, conserving water and fostering a healthier ecosystem. Native plants, adapted to local climates, thrive with rainwater, attracting diverse wildlife and contributing to the preservation of local biodiversity.

4. **Soil Health and Regenerative Agriculture**

   Rainwater harvesting plays a crucial role in promoting soil health and regenerative agriculture. By capturing rainwater and directing it to agricultural fields, farmers reduce their dependence on unsustainable water sources and adopt more environmentally friendly irrigation practices.

   The replenishment of soil moisture through rainwater harvesting contributes to improved soil structure and fertility. This, in turn, supports sustainable farming practices, reduces soil erosion, and enhances the overall resilience of agricultural ecosystems.

5. **Mitigating Flood Risks and Stormwater Management**

   In urban areas, heavy rainfall often leads to flooding and strain on stormwater management systems. Rainwater harvesting acts as a natural solution to mitigate these risks by reducing surface runoff.

When rainwater is harvested and used on-site, less water enters stormwater drains, lowering the risk of flooding. Additionally, the process of collecting rainwater helps filter out impurities, reducing the burden on stormwater management infrastructure and improving water quality.

6. Energy and Cost Savings

Rainwater harvesting leads to energy and cost savings. Traditional water supply systems, which involve pumping and treating water for distribution, consume significant amounts of energy. Using locally harvested rainwater decreases the demand for centralized water supply, resulting in reduced energy consumption and lower utility costs.

7. Drought Preparedness and Climate Resilience

As climate change leads to more frequent and severe droughts, rainwater harvesting becomes a vital tool for drought preparedness and climate resilience. By capturing rainwater during times of plenty, communities can build reservoirs for use during drier periods.

This proactive approach to water management contributes to climate resilience, ensuring a more reliable water supply in the face of changing weather patterns. Rainwater harvesting acts as a buffer against the uncertainties associated with climate change.

As individuals, communities, and societies, the commitment to rainwater harvesting is a commitment to a sustainable future. It is an acknowledgment that every drop saved today is a gift to the generations of tomorrow. The journey toward sustainability is a collective endeavor where each raindrop harvested becomes a symbol of hope, resilience, and the promise of a thriving planet for all.

Part 2: Water Survival Guide

Mastering the Art of Finding, Collecting, Treating, Storing, and Thriving Off the Grid with Water for Emergencies and Outdoor Adventures

Introduction

Water is the giver of life from which everything emerges. Therefore, it is one of the first resources you need when going off-grid or in an outdoor survival emergency. Gathering, treating, and storing your water are all skills you need to master for a truly self-sustained existence, free from the shackles of societal systems that aim to hold you back. As a survivalist, prepper, outdoors person, or an off-grid homesteader, you can benefit from the mountain of enlightening information you are about to explore.

When you've finished this book, you'll have an in-depth education on the numerous ways you can gather and clean water. Moreover, you will have focused your survival eye needed to transform any space into a hub to thrive. By understanding natural processes and different biomes, you can tap into the flow of the elements to manifest an abundance of this life-giving and sustaining liquid. Once you internalize some basic principles and techniques, your water survival skills will be fine-tuned and elite. The mix of theoretical knowledge and practical advice gives you a complete view of the intricacies of working with water and the many considerations people forget when they run their bath water.

With a new-found respect for water and its uses, you will unlock ways to think about sustainability. This text explores hydro geography and how the land can be used to your advantage when seeking this essential natural resource. You will also learn how to purify water and make sure that you are kept healthy and strong. Furthermore, you will obtain the knowledge to collect rain and dew in a multitude of different ways. You will even learn how to work with water in the tundra by debunking some of the myths of snow and ice melting while learning the best ways to gain water in these freezing environments.

With a focus on conservation, you will be shown how to gather water in a permanent homestead or when you are on the move. You will be equipped with the skill to survive in settings ranging from water-scarce to water-abundant by learning how to transport and store it for long-term use. Prepare to unfold the multifaceted reality of a resource often overlooked and undervalued as you walk the path of water survival outdoors and thrive off the grid.

Chapter 1: Water: The Essence of Life

While exploring the intricacies of life, you encounter a pivotal element threading through the essence of existence – water. It's a valuable resource embodying the life force that sustains and nurtures all living beings. Not only that, but the Earth's ecosystem also is majorly regulated by water, creating striking and serene landscapes and areas where living beings can inhabit and thrive.

From conception onward, water orchestrates the symphony of human life. Although the human body is 60% water, several vital organs like the heart, brain, and lungs comprise 75 to 80% water. Look around, and you'll find water as the fundamental building block on which life flourishes. It's interlaced with the fabric of human civilization. Ancient cultures revered seas, oceans, and lakes as sacred entities as water promoted sustenance and expansion. To the present day, you can find scriptures and accounts of historians from several civilizations depicting The Nile, Ganges, Tigris, and Euphrates as cradles for great civilizations, shaping the course of history and facilitating the growth of human societies.

Water is the elixir of life.
https://pixabay.com/photos/water-droplets-dewdrops-water-955929/

 Water's subtle yet profound role in daily life is mostly overlooked in this technologically driven era where humans are carrying out space explorations on an exponential level and driving industries through cutting-edge technologies. It quenches thirst, keeps the body hydrated, helps detoxify, irrigates fields, generates power, and fuels industries. Despite its omnipresence, the precious nature of water often escapes everyone's notice until its scarcity looms on the horizon.

 Water itself is the elixir of life, as most living beings depend on this precious source to stay alive. It isn't just something everyone uses. It's the critical ingredient that keeps life ticking on Earth. From tiny bugs to big animals, water is required for everyone. It's like the builder that makes homes for all the different plants and animals.

 Right from the very start of your life, water has been there with you. It's like a cozy bed in a mommy's tummy, helping you grow. Water helps people stay healthy and strong. It's not just about the body. Water shapes how people live together, making communities and traditions possible. Think about the rivers and oceans – they're like big lifelines for human beings. Human ancestors knew this, too. They set up homes close to rivers because they knew rivers were like nature's supermarkets, providing food and water. Even now, in the modern world, water is a critical resource. It helps grow food, keeps living beings hydrated, and powers most industrial processes that shape this era.

Life's Essential Drink

Water is the universal drink for all living things. Plants, animals, and humans need water to survive. It's the basic stuff that keeps your body working. Without water, body functions will deteriorate and ultimately cease to exist.

Keeping Things Cool

Water is nature's air conditioner when the weather gets too hot. It helps cool down the Earth and keeps living things from overheating. It's essential to maintain the ecosystem through natural processes like rainfall.

Growing and Thriving

For plants, water is their food. It helps them grow, make their own food through photosynthesis, and stay green and healthy. Likewise, every other living being needs water to thrive, contributing to a food web from which all animal species benefit.

Quenching Thirst

No matter what beverage, juice, or sugary drink you consume, water is the only source that quenches your thirst. Although its chemical composition is simple, the intestines can easily absorb it, reaching every body part to become body fluids like blood and facilitate metabolic processes.

Essential for Hygiene

Water is not just for drinking. It's also the best liquid for washing. Humans use it to take baths, keep homes tidy, and keep the dishes sparkling. Replacing water as a cleaning substance is nearly impossible. Without it, maintaining hygiene and cleanliness will become nearly impossible.

Powering the World

Beyond keeping life thriving on this green planet, water is used worldwide to power hydroelectric dams. Although solar and wind electricity generation is making waves in terms of clean energy, hydroelectric power generation is still the primary source of electricity that powers homes and runs the gadgets you love to use.

As water is necessary for life itself, you must use it carefully. Pollution and wasting water can harm the environment and make it challenging for everyone, including humans, to have enough clean water. Massive urbanization, the use of clean water for industrial purposes, and the addition of pollutants into streams and rivers are contributing to water waste.

Water is not just a drink. It's a lifeline for all living things. For humans, it goes beyond survival, and it's part of what makes your lives healthy, comfortable, and meaningful. Taking care of water is caring for yourselves and the planet you call home.

The Human Body and Water

Cell Function and Structure

Water is the building material for your cells. It helps form the structure of your cells and ensures they function correctly. It allows the cells to have the right shape, down to the molecular level. Water maintains the cell turgidity and makes body fluids that maintain cell structure and function.

Digestive Dynamo

Why do most people say you should drink water before meals? It's because water is a crucial player in your digestion. It breaks down your food, making it easier for your body to absorb all the good stuff and eliminate the stuff you don't need. Water is absorbed easily through the intestines and is essential to maintain the body's water weight.

Temperature Tamer

You start to sweat when you feel a rise in body temperature after playing sports or on a sunny day. It is your body's way of cooling down, and water is the main ingredient in sweat. So, in a way, water is your built-in air conditioner.

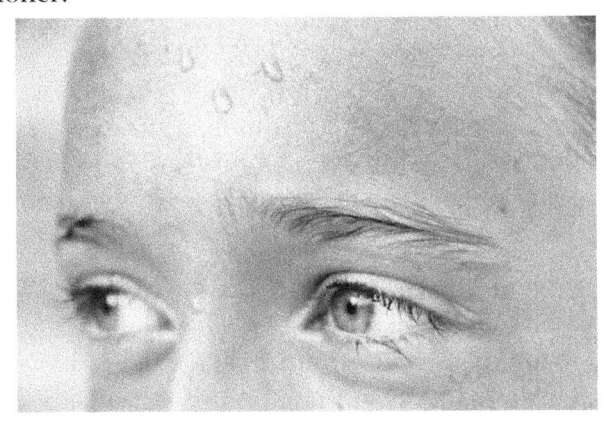

Water is the main ingredient in sweat.
Photo by Hans Reniers on Unsplash https://unsplash.com/photos/persons-eyes-looking-on-left-side-mE6e5-5jLu8

Joint Lubrication

Have you ever thought about how your knees or elbows move so smoothly? Thank water for that. It lubricates your joints, making it easier for them to bend and flex without any squeaky sounds. After absorption in the intestines, water goes through a series of processes, forming the synovial fluid, which acts as a cushion and a shock absorber. It reduces friction in the joints and surrounding cartilage while moving, keeping the bones from rubbing together.

Kidney's Cleaning Crew

Your kidneys are your body's janitor, cleaning up the waste and extra stuff you don't need. Water is their trusty sidekick, helping flush out all the things your body doesn't want to hang around. Most of these unwanted substances enter the body through food consumption, and some chemicals are released as a by-product of processes occurring within the body. All these toxins are flushed through the kidneys in the form of urine.

Blood Flow Booster

Your blood is mostly water, and water helps the blood flow smoothly through your veins and arteries. Good blood flow means all body parts get the necessary oxygen and nutrients. Without adequate water levels, the vessels fail to maintain a consistent diameter, affecting blood flow and reducing the flow of nutrients throughout the body. This increases the chance of blood clots, which can trigger other chronic conditions.

Muscle Function

Water is the fuel that powers your muscles. When you're active, your muscles use water to function correctly. So, whether you're playing soccer or doing cartwheels, water is your secret energy source. Besides enabling free movement, water also promotes muscle gain as it acts as a carrier, delivering materials required for essential proteins and glycogen synthesis. The body's main energy sources are proteins, glycogen, and related materials.

Skin Savior

Ever notice how your skin looks better when you drink enough water? That's because water keeps your skin hydrated, making it look healthy and glowing. Staying hydrated locks in the moisture, which aids in maintaining skin elasticity. This increased elasticity prevents the skin from sagging, reduces wrinkles, and prevents fine lines from developing. Lastly, water also supports the production of collagen, a substance produced by the body that promotes the development of new skin cells by replacing dead skin cells. It's the best beauty treatment, and it's free.

Brain Buffer

Even if your brain needs water, it helps you think clearly, concentrate, and stay alert. Water facilitates neural connectivity, allowing the brain to communicate effectively. It also clears out the toxin build-up within the brain, optimizing its function. So, if you're ever feeling a bit foggy, a glass of water might be just what your brain needs.

Water is your body's MVP (Most Valuable Player) of bodily functions. It's not just about quenching your thirst; it's the unsung hero behind the scenes, ensuring everything in your body runs smoothly and efficiently. In the realm of essential needs, its scarcity casts a looming shadow on human health. The critical importance of water becomes starkly evident when you consider how brief your survival would be without it.

Life without Water

Threat to Survival

Humans can survive without food for a relatively extended period, but the same cannot be said for water. The body's dependence on water is immediate and critical, with survival jeopardized after just a few days without this precious resource.

Dehydration's Stealthy Onset

The onset of dehydration, a consequence of insufficient water intake, is swift. The human body loses water through daily activities, including breathing, sweating, and waste elimination. These losses escalate in a water scarcity situation, leading to a rapid decline in bodily functions. Over time, these functions will halt, creating a life-threatening situation.

Impact on Physical Health

As discussed earlier, water scarcity contributes to a cascade of health issues. Dehydration impairs the body's ability to regulate temperature, increasing the risk of heat-related illnesses. It also affects the circulatory system, reducing blood volume and increasing potential cardiovascular complications. Furthermore, body fluids like cerebrospinal fluid (essential for the brain) and lymph (required for immune system functioning) won't be produced in the required amounts, compromising cognitive function and weakening the immune defenses. As the body's metabolism slows down and the immune defenses weaken, it opens a window for chronic illnesses, infectious diseases, and medical conditions to affect the body.

Water-Related Diseases

The scarcity of clean water often forces communities to rely on contaminated sources. This, in turn, leads to waterborne diseases, posing severe health threats to populations facing water shortages. Common waterborne diseases include cholera, dysentery, giardiasis, typhoid, scabies, amebiasis, hepatitis, and parasitic infections if the water source is contaminated.

Mental and Cognitive Impacts

Dehydration extends beyond physical health, impacting cognitive function. Even mild dehydration can result in decreased alertness, difficulty concentrating, and impaired short-term memory. Prolonged water scarcity exacerbates these effects, hindering overall mental well-being.

Vulnerability of Specific Groups

Infants, the elderly, and individuals with pre-existing health conditions are particularly vulnerable to the health impacts of water scarcity. Their physiological fragility amplifies the risks associated with dehydration and waterborne diseases.

Aggravation of Existing Health Disparities

Water scarcity exacerbates existing health disparities, disproportionately affecting communities with limited access to clean water.

Socioeconomic Ramifications

Beyond the immediate health concerns, water scarcity amplifies socioeconomic challenges, such as increased healthcare costs, reduced productivity, and a strain on communities. In the face of these issues, the detrimental effects of water scarcity on human health are not just hypothetical scenarios but pressing realities. Recognizing the urgency of ensuring equitable access to clean water becomes paramount to safeguarding human health and well-being.

Food and Water in Life

Survival Imperatives

In situations where water is limited, it immediately impacts the body's fundamental physiological processes, like hydration and nourishment, which are crucial for sustaining life. The absence or scarcity of water poses an imminent threat, making it crucial to find and secure access to water for survival. This water scarcity affects humans and the entire food chain, from microorganisms and plants to animals. Virtually every other species on this planet has a critical role in maintaining this food chain. For example, birds and flying insects are crucial for pollination, and certain species of bacteria and fungi are responsible for decomposing dead organic matter, giving back to the Earth, and so on. When water scarcity peaks, there's a disruption in this complex chain of events, resulting in survival issues for every living being within the ecosystem.

Physiological Vulnerability

As you already know, the human body is an intricate system with tightly interlinked physiological processes. Deprivation of one essential resource, whether for hydration or nourishment, rapidly disrupts these processes. For example, lack of hydration affects temperature regulation, circulation, and immune function. When one of these life-fueling processes within the body doesn't function adequately, it affects every other body function. Timely provision of water and food is essential to keep the body up and running in optimum condition.

Disease Susceptibility

The scarcity of clean resources heightens the risk of disease. Contaminated or insufficient resources can spread waterborne illnesses, malnutrition, and other health issues. This susceptibility to disease threatens public health, necessitating urgent measures to secure access to clean and sufficient resources.

Vulnerability of Specific Groups

Infants, the elderly, or individuals with pre-existing health conditions are particularly vulnerable to the immediate health impacts of resource scarcity. The unique physiological needs of the body and the processes involved get disrupted without the provision of water. If water and food scarcity persist, it can further deteriorate the metabolic processes and even lead to life-threatening situations.

Likewise, underprivileged communities in hot and arid climates struggle to access water and food. This inadequate access causes health to deteriorate rapidly, further burdening healthcare costs. Inadequate provision or scarcity of water and food takes a massive toll on communities like this as they are already struggling to make ends meet.

Impact on Habitat

Likewise, water scarcity also affects other sources of food. For example, if streams in a region dry up and the groundwater levels plummet, the habitat changes drastically, affecting all living beings. From microorganisms to plants, wildlife, and humans, the area will slowly become inhabitable, eliminating virtually every form of life.

Environmental Adaptation Challenges

Both human and natural systems face challenges in adapting to abrupt changes in resource availability. Ecosystems may struggle to cope with shifts in water availability or changes in agricultural practices. Human systems also need to change quickly when resource supply changes.

This detailed exploration highlights the urgency of securing access to critical resources for survival, the vulnerability of physiological processes, the increased risk of diseases, the unique susceptibility of specific populations, the societal and economic repercussions, and the challenges associated with environmental adaptation. Recognizing and addressing these aspects is crucial for effective resource management and the well-being of individuals and communities.

As you reach the culmination of our exploration, the profound significance of water as synonymous with life becomes undeniably clear. The journey through the intricate web of water's roles – from sustaining physiological processes to fostering societal well-being – reveals why this precious resource is often reverently termed 'life-giving.'

Water isn't just a resource; it's the universal elixir that breathes life into everything. Every living creature depends on water for survival and vitality, from the tiniest creatures to the tallest trees.

Throughout the chapter, you've read how water is more than just a drink. It's the foundation of human life, from the earliest days of development to the sustenance of communities and the building blocks of civilizations. It's the nourishing force that shapes the very course of your life. The essence of water as life is not just a metaphor but a fundamental truth. Its deprivation brings forth consequences that resonate from your individual experience to the societal level. As a steward of this precious resource, you must cherish and safeguard water, not just as an environmental duty but as a commitment to the very essence of your life. Through understanding, appreciation, and collective action, you can preserve the vitality of water for generations to come.

In the coming chapters, you'll uncover everything related to water, the tools required to harvest water naturally and serve as a comprehensive water survival guide to follow that can keep you going in unexpected situations and the harshest conditions.

Chapter 2: Hydro-Geography

If you've ever traveled alone in remote areas, you've likely had to find water by the second or third day, at the latest. Finding water isn't usually hard unless you're in an arid area. Typically, you just need to check the quality and treat visible water from big rivers or lakes. The key to finding water in nature is paying attention. When you're in an emergency, the initial reaction is often panic and stress. However, getting hysterical makes things worse, so it's crucial that you stay calm. Being prepared helps you to keep a cool head.

Don't be too worried if you don't see obvious water sources like rivers, lakes, streams, or puddles nearby. Take a few deep breaths. Look out for the signs that nature provides. Since every living thing needs water, the presence of any form of life indicates that there may be usable water close by.

Any form of life indicates that there may be usable water close by.
https://pixabay.com/photos/raindrops-sheets-ladybug-574971/

Understanding Terrain Features for Finding Water

In arid regions, readily available drinking water is rare. Potential water sources can sometimes only be noticed when physically stepping into them. Additionally, crucial signs such as soil layers, indicator plants, and animals can often be identified only in close proximity to a water source, making observation important.

To find promising areas, it's essential to consider the geological characteristics of the environment. Water flows downhill and collects or goes underground at the lowest point. Surveying the surroundings, you can easily identify signs of existing surface water or subterranean water levels.

When examining geological features, it's helpful to know that rock, confining soil layers, and similar structures continue underground in the same pattern as they appear above ground. Disregard the soil layers above a confining layer and visualize how it extends below them.

For instance, if you estimate an intersection of two rocky mountainsides to be beneath some erosion rubble between the slopes, you can mark this point as a potential runoff point. Remember that water may not necessarily break through to the surface but may flow off in the sediment many feet below. Despite this, these natural reservoirs have a significant advantage as they are not affected by evaporation due to the lack of sunlight and can hold water for decades after the last rains.

In certain situations, water accumulates above the water table when it can't drain away quickly. Generally, the water level of open bodies indicates the local groundwater level, lying between a confining soil layer (aquiclude) and surface water level. Trenches in a mountainside suggest the course water can take, indicating the location of water runoff either above or below ground. Canyons or gorges signify water erosion following heavy precipitation events. Follow the dry riverbed until you come across a barrier; if there's a subterranean water reservoir, it might be there. A mountain saddle without natural runoff may contain an open mountain lake or water reservoir under the sediment, with very slow runoff through the bedrock.

Overlapping cliffs with deep valleys represent potential points for a surface river or pond. In a depression in an extensive plain, if there's any subterranean water, it's likely to be found at the lowest point of the depression.

Detecting Water Presence: Animal Tracks and Indicator Plants

Tracking wild animals is often a reliable method for finding water in a natural environment. However, some mammals, especially those in arid regions, have adapted to alternative water sources or can endure long periods without water, making animal tracks less dependable. An extreme example is the desert-dwelling kangaroo rat, which recycles condensed water from its breath in its nasal mucous membrane. Larger mammals may travel great distances to reach the nearest watering hole.

Stories circulate about insects and their relationship to water, but these should be approached with caution. While bees and mosquitoes are known to depend on water, this is mainly true for species in moderate and humid climates. Wind can blow individual animals or swarms into water-free zones where they can survive for weeks or months without water.

Indicator Plants and Animals

- **Ferns, Mosses, and Horsetails:** These plants depend on water for their development, and their widespread global presence makes them valuable indicators of moist or intermittently liquid soil. When you come across areas abundant with ferns, mosses, and horsetails, it signals the likelihood of nearby water sources.
- **Fast-Growing Water Meadow Trees (e.g., poplar, willow):** The characteristics of fast-growing water meadow trees, such as poplar and willow, reveal important information about the surrounding environment. Their shallow roots and preference for temperate zones make them reliable indicators, suggesting the presence of moist soil or surface water in the vicinity.
- **Pandanus:** Found on beaches and prevalent in Oceania's rivers, lakes, marshlands, and freshwater lenses, the presence of pandanus plants is a significant clue. These plants serve as indicators, particularly on beaches, hinting at the potential existence of freshwater lenses beneath the surface.
- **Taro and other (large) Araceae:** The large foliage of taro and similar plants is a clear indicator of moisture in the soil. Commonly found in Southeast Asia near rivers, lakes, and marshlands, their presence signals areas with consistently moist conditions, providing valuable information for water seekers.

- **Reeds, Bulrushes:** Thriving in eutrophic waters globally, reeds and bulrushes are reliable indicators of permanently moist soil. Their presence signifies areas with a high likelihood of liquid water, offering essential clues for those searching for water sources.
- **Tree Ferns**: Surviving only short dry spells, tree ferns found in tropical and subtropical regions indicate the presence of consistently moist soil. When you encounter these ferns, it suggests a local environment with a stable water supply.
- **Rats:** With a global distribution, rats are often found near slow-flowing rivers, lakes, and canals. While not directly indicating the presence of water, their proximity to water bodies can be a relevant consideration, especially when evaluating the overall ecosystem for survival needs.
- **(Water) Snails:** The larger species of snails, equipped with an epiphragm, are often found in moist soil with a high water table worldwide. Their presence indicates areas with accessible water sources, as these snails thrive in environments where water is readily available.
- **Cockatoos, Pigeons, Macaws, etc.:** Mass gatherings of birds like cockatoos, pigeons, and macaws are audible and visible signs of water from long distances. These birds are commonly found near rivers, lakes, and marshlands globally, making their presence a reliable indicator of nearby water sources.

In the Australian outback, bees are not a reliable indicator for water proximity. Many insects and arachnids can survive with rare, microscopically small dew drops for their development stages and water supply.

Certain creatures are particularly useful water indicators, either in liquid form or contained in the soil. These include stationary organisms like plants and highly mobile animals that congregate in large numbers, such as birds.

Indicators for Moist Soil

- In temperate regions, reeds and rush pads are vital water indicators growing in habitats with very moist soils. Their presence usually suggests the availability of liquid water nearby.

- Globally, reeds or cattails are easily identifiable by their long, broad leaves and indicate wet to marshy soils. They can also serve as a water substitute.
- Around the equator in Asia and Africa, taro and other large-leaved Araceae are significant water indicators, thriving in wet conditions.
- Tropical tree ferns, unable to endure long dry spells, settle in the permanently moist ground on hillsides and valleys, indicating subterranean water flows in mountainous regions.
- Even desert-dwelling palm trees, needing water due to fluctuating water tables, can be found in areas without surface water. Their presence suggests the likelihood of water, even in seemingly barren places.

Identifying Potential Groundwater Sources

1. Springs

The most straightforward way to access water is through springs, where subterranean water naturally emerges above ground. Springs can take various forms, such as "rheocrene," where water trickles or gushes out. Often, springs flow into ponds or marshes when the groundwater table is above ground level. Seepage springs are smaller bodies of water without a clear inlet or outlet lying on confining soil layers. Recognizing likely locations involves understanding the geological structures of the landscape.

Indicators of Reliable Springs

A reliable spring may have a relatively low water temperature (around 41 to 50°F [5 to 10°C] in temperate zones, approximately 68°F [20°C] in the tropics) compared to the ambient temperature. Torrents may carry more water beneath the surface than visible, especially in areas with loose soil. A confining soil layer beneath loose stones can store and discharge water in lake- or river-like structures without being apparent above ground.

2. Wells

Wells are artificial attempts to reach the groundwater table. While typical wells may not need detailed instructions, it's crucial to remember that wells are often private property, requiring permission from the owner before using the water. The term "well" encompasses any artificial structure reaching the groundwater table, including simple holes dug to a depth of one to two meters (three to seven feet).

Wells are built to reach groundwater.
PePeEfe, CC BY-SA 4.0 <https://creativecommons.org/licenses/by-sa/4.0>, via Wikimedia Commons: https://commons.wikimedia.org/wiki/File:Water_well_-_Coste%C8%99ti_-_Romania.jpg

Infiltration Wells

Digging may be worthwhile in regions where geological features or other reliable factors suggest the water table is close to the surface. In subtropical regions, water might not be extractable as a liquid but instead collected as moist earth or mud from the unsaturated zone using a suction well. A high groundwater table is indicated by the presence of nearby bodies of water, even if polluted. When digging, pay attention to foul smells from the soil layers, as they could mean the water is polluted.

Important Considerations

- Wells on private property require permission before use.
- Infiltration wells make sense in areas where geological features or reliable indicators suggest a close water table.
- The proximity of a body of water helps estimate the groundwater level.
- Foul smells during digging may indicate polluted water, requiring adjustments to the well location.

Understanding these aspects can improve your ability to locate and access groundwater in a range of terrains and conditions.

Suction Wells

When the sediment lacks enough moisture for liquid water extraction, constructing a suction well is worth considering. Cover the pit's bottom with plant material or clothing fabric, placing a cane, plastic pipe, or similar object inside. Refill the pit with tightly packed soil, allowing liquid water to be sucked into the material and, subsequently, through the cane or tube into your mouth. If this method proves unsuccessful, swallowing moist earth directly or squeezing it through fabric may be necessary. In extreme cases, you may have to distill the water.

Freshwater Lenses

While the term "freshwater lens" may evoke images of underwater photography, it actually denotes a vital phenomenon that safeguards drinking water on islands and secluded beaches. To grasp the dynamics of this freshwater source, you need to delve into the actual concept of density. In essence, dissolving a soluble substance in water leads to the breakdown of molecules or charged ions, dispersing within the water molecules and altering their density. Picture making an impromptu water filter from a combination of coarse gravel and fine sand. The addition of sand doubles the weight but barely affects the volume, showcasing the impact on density.

Swap the sand for table salt, and you'll have a different outcome. The salt dissolves, splitting into Na^+ ions and Cl^- ions (dissociation), increasing density. This disparity in density explains why saltwater outweighs freshwater. A liter of freshwater weighs around 1 kilogram, while seawater of the same volume is about 30 grams heavier. Similar to how Styrofoam floats while lead sinks, freshwater (lower density) rests on top of saltwater. This phenomenon, known as a freshwater lens or Ghyben-Herzberg lens, forms in porous soils like sand, especially where they meet the sea, given occasional rainfall.

Unlike conventional groundwater, freshwater lenses don't rely on impermeable layers; the saltwater beneath the island acts as a confining layer. Notably, freshwater lens levels are higher than sea level. When seeking fresh water on the beach, avoid lower sections and explore the island plateau's highest points for pools or water indicators.

True freshwater lenses may be absent on smaller islands, replaced by brackish water. Salinity assessment becomes crucial in such cases.

Meanwhile, submarine freshwater sources on mainland beaches, driven by subterranean rivers, are discernible through geological structures like mountainous trenches or depressions sloping toward the sea.

Collecting water from these sources resembles freshwater lens methods, but it's closer to the beach due to constant interior flux. Contrary to myths, sand doesn't "filter" salt, and opaque streaks in seawater indicate submarine springs. Inland freshwater sources emptying into the sea are prerequisites for beach freshwater. If inland water isn't present, anything dug near the sea is likely to be brackish or salty.

Movement of Water Across Landscapes

Water movement across landscapes is a dynamic process influenced by various factors. Understanding these patterns will make finding water sources in different terrains much easier.

Surface Water Flow

Surface water flow refers to water movement across the Earth's surface. This can occur through various channels, including rivers, streams, and runoff from precipitation. The direction and speed of the surface water flow are influenced by the land's topography. Water follows the path of least resistance, often carving out valleys and channels as it flows downhill.

Recognizing surface water flow patterns involves observing the land's contours and identifying potential watercourses. Understanding the interconnectedness of surface water systems aids in predicting where water is likely to accumulate or flow during rainfall.

Underground Water Flow

Underground water movement, also known as subsurface flow, occurs beneath the Earth's surface within aquifers and permeable rock formations. This subsurface water flow is critical for sustaining groundwater levels and surface water features. Identifying underground water movement involves assessing geological formations and the characteristics of aquifers.

Springs, where groundwater emerges at the surface, are indicators of subsurface water flow. Recognizing the relationship between surface and underground water movement provides valuable insights into overall water availability in a given area.

How Landscape Features Impact Water Movement

Slopes and Gradients

The slope and gradient of the land significantly influence the direction and speed of water movement. Water naturally flows downhill, following the contours of the terrain. Understanding slope dynamics helps predict where water may accumulate or form channels.

Steep slopes contribute to rapid surface water flow, potentially leading to erosion and the formation of valleys. Conversely, gentle slopes may result in slower water movement, allowing for water retention and soil moisture.

Soil Permeability

Soil permeability, or the ability of soil to transmit water, is a crucial factor affecting water movement. Different soil types exhibit varying permeabilities, influencing how water is absorbed, retained, or drained. Sandy soils, for instance, typically have high permeability, allowing water to pass through quickly.

Recognizing soil permeability helps in assessing the likelihood of water retention in the soil and the potential for groundwater recharge. It also aids in understanding how vegetation interacts with soil moisture, providing additional clues for water detection.

Common Misconceptions and Myths

Finding water can be tricky, and plenty of myths may lead you down the wrong path. This section is about setting the record straight on some common misunderstandings and ensuring you know the right way to find water in different places.

Debunking Myths for Different Terrains

- **Deserts:** You may have heard that digging in low spots in the desert will reveal water. Well, it's not that simple. Water doesn't always collect in low areas. Understanding the local geography, like wadis (dry riverbeds) or underground water sources, is more reliable.
- **Forests:** It's a mistake to assume that all parts of a forest have easy access to water just because it's a lush environment. Reliable

water indicators, like animal tracks leading to water, are more trustworthy than general assumptions.

Evidence-Based Approaches

- **Smart Techniques:** Forget about mystical methods or using divining rods. Trusting scientific, evidence-based techniques is far more reliable. Knowing your terrain, understanding hydrological principles, and recognizing reliable visual clues are much more reliable.
- **Visual Clues**: Instead of relying on rumors, it's important to be able to identify visual clues in the landscape. Knowing how to spot potential indicators, like specific vegetation patterns, geological formations, and animal behavior, is key to finding water.
- **Understanding the Terrain:** Skip the guesswork and go for a systematic analysis of the terrain. Knowing how topography affects water movement gives you a much better chance of predicting where to find water.
- **Hydrological Know-How:** Get familiar with basic hydrology principles. That means understanding how surface water flows, how groundwater behaves, and how different soils handle water. This knowledge is your secret weapon.

Adapting to Different Environments

- **Tailor Your Approach:** Not all terrains are identical, so your approach shouldn't be either. Customize your methods based on where you are. What works in one place may not work in another.
- **Consider the Climate:** Think about how climate affects water availability. In dry areas, certain plants might be reliable indicators, while in wetter places, you may need to understand how water moves beneath the surface.

In a nutshell, let go of the myths and focus on solid, evidence-based methods. By understanding the unique challenges of different terrains and relying on science, you'll be much more successful in finding water where you need it. It's not just about survival—it's about being smart and informed in the great outdoors.

Chapter 3: Rainwater and Dew Collection

One of the first considerations you must make when pursuing a life off-grid is how you will get water. There isn't always the luxury of having a big body of water like a lake or river nearby, so rain and dew collection may be the only options for sustainable water collection. Any survivalist must have a constant supply of water, so collecting rain may arguably be the easiest way to achieve self-sufficiency.

When you open your faucet and water flows out, it has already gone through multiple processing steps to be safe for human use. If you collect your water, the responsibility for making sure your water is usable now falls fully onto your shoulders. Therefore, it is essential to learn the best methods of collecting water and how to process the life-sustaining liquid afterward.

Rainwater and dew.
https://pixabay.com/photos/raindrop-drops-car-roof-heaven-4929166/

By understanding the natural processes of nature and how you can use them to your advantage, you will be equipped with the knowledge to set up workable survival systems. Furthermore, you will be informed about the essential safety protocols needed to sustain your health and use water responsibly. From weather patterns and ecosystems to microbes and filtration, the world of rainwater and dew collecting is demystified with practical tips and credible information.

The Process of Rain and Dew Formation

To harvest dew or rain, you must be in the right environment and understand the functioning of natural processes. If you are in a drier area, you may need more storage, whereas if you are in a tropical climate where rain is abundant, the potential of flooding is going to be an issue to consider when collecting rainwater. If you are not near a natural water source, or the water body you have access to is contaminated, the best option you have is collecting rain and dew.

Rainfall is part of the water cycle, which has four main parts. The first section of the water cycle is evaporation. The sun heats surface water, speeding up the molecules so much that the liquid water turns into water vapor. The second part of the water cycle is when the vapor reaches the sky. The atmosphere cools the vapor, condensing it into clouds. When the clouds are heavy and saturated, the third step of the water cycle occurs, which is precipitation. Water vapor condenses back into liquid, causing it to fall back down to the earth. The last part of the cycle is the formation of groundwater, which is the best time to collect it. The cycle constantly repeats. You can collect rainwater at any stage in the cycle. A mesh net can be used to catch evaporating water. Fog catches can be used to gather low clouds in mountainous regions. The most apparent and efficient way to collect rainwater is to catch surface runoff.

Dew forms in a slightly different way than rain, but many of the same processes are still involved. Dew is a result of condensation, which is the process where gas transforms into liquid. At night or in the early morning, the surface of objects cools down because the sun is no longer hitting them directly. When a surface or object gets cold enough, the air around it will also cool down. Warm air holds water vapor better than cold air. The cold air condenses the water vapor, which causes small droplets to form. The dew point is the sweet spot where dew can develop. The formation of dew depends on numerous environmental factors like your geographical location, as well as temperature and wind. Dew forms better in humid regions like tropical rainforests. In deserts, the conditions are not suitable for a lot of dew to form because the air is too dry. Therefore, all you need to collect dew is the right surface and an appropriate region that has all the needed environmental conditions.

Rain and Dew Collection Methods

Since ancient times, humans have settled near water sources. If you study any civilization, you will find that they sprung forth from one centralized place that either had a river, spring, or lake. The Ancient Egyptian civilization was birthed from the rich agricultural land near the Nile, and Rome rose from a marshy swampland in Italy. When setting up an off-the-grid homestead, or if you are in a survival situation, sometimes the only option you have is to collect rainwater.

Rain and dew collection are cheap and convenient options for obtaining water, which can be set up quickly. There are some downsides of rain collecting, like seasonal rainfalls and droughts. However, with some luck and the proper setup, you could have a rain and dew water system that runs throughout the year. The easiest setup for collecting water is using tanks to collect surface runoff. Plastic water tanks created for this use are available on the market. The water tank will be set up near a roof or somewhere rainwater can flow off. Gutters are connected to the tank to collect the water. It's then stored in the tanks, which can be accessed when needed. The beauty of this method is that you can use the same principles to create cheaper options by using recycled materials. However, you should always ensure that the container you store your water in does not house harmful contaminants.

Smaller barrels can also be used as a cheaper option. However, the amount of water you can collect will drastically decrease compared to the much bigger water tank. The upside about using barrels is that they will be easier for you to move. Full water tanks need machinery to be relocated and will take far more effort. Setting up a few barrels around your home near the gutters can help you meet some of your basic water needs, like washing or watering the garden, but it is unlikely you can go fully off the grid by relying solely on this method.

A gabion is a more permanent way to collect rainwater.
No machine-readable author provided. Zimbres assumed (based on copyright claims). CC BY-SA 2.5 <https://creativecommons.org/licenses/by-sa/2.5>, via Wikimedia Commons: https://commons.wikimedia.org/wiki/File:Gabion1.jpg

A more permanent and labor-intensive way to collect rainwater is a gabion. This basic dam is made up of rocks placed within a wired boundary wall. A concrete mixture is then placed over the rocks to seal the gaps. You can collect a lot of rainwater depending on how big your gabion is. Since the water will be still, you must be precise with your disinfection and cleaning processes. The materials that are used are relatively low-cost. The wire mesh for the walls can be purchased at any local hardware store, and the rocks can also be bought or even collected depending on the land you have access to.

Swales and canals will be very useful if your rainwater is to be gathered for farming purposes. Swales are shallow ditches strategically dug to channel water from large surfaces downhill toward crops. If done correctly, this simple irrigation system can save you thousands of dollars per year in farming costs for any subsistence off-grid community or homestead. Many rural communities in Africa and Asia use swales because their livelihoods are directly linked to their success with the land.

Dew collection works best in areas with high humidity. Harvesting dew is an often-overlooked way to get water for self-sufficient living. Methods of dew collection can sometimes seem crude, but they are effective. The first simple way to harvest water from dew is to collect it from the grass.

First, you will need a storage container. Then, you need an absorbent cloth made from wool, cotton, or a similar compound of synthetic materials. When the dew has settled in the early morning, lay your cloth on the grass to absorb the water. Once the cloth is heavy with moisture, wring it out into your water-collecting container. Repeat this process until you have collected all the water that you need. This method can also be used for other plants and leaves, but ensure your flora is non-toxic before applying this technique.

A waterproof tarp can be used for dew collection.
ModalPeak, CC BY-SA 4.0 <https://creativecommons.org/licenses/by-sa/4.0>, via Wikimedia Commons: https://commons.wikimedia.org/wiki/File:Forester_Tent_Tarp_and_Poles_(6262CE06).jpg

A waterproof tarp can also be used for dew collection. When deciding upon any survival prepping or off-the-grid lifestyle, a central principle is to use what you have. Many people have tarps lying in their garages or basements. For this water-harvesting method, you could also use raincoats or an old tent as a makeshift tarp. Find a slope or construct one wherever you have some space out in nature or on your land. Use bricks and soil to pin the tarp down. Lay the tarp in a triangle shape on the ground, folding the sides over your bricks and soil and securing the material to the ground. Create a funnel shape at the end of the tarp. Make sure that soil is packed underneath your tarp to cool the material at night. Place a container at the end of the funnel you have created and allow the dew droplets to flow slowly into it. Gravity helps you with this method. That's why you need to construct the contraption on a slope.

Traditional Approaches and Modern Innovations

Ancient cultures understood the value of rainwater collection. Thus, various systems were developed around the world to harvest rain. Today, rainwater is still collected in big cities through a complex network of gutters and greywater constructions that channel water into dams and through processing plants. Traditional rainwater harvesting methods are cruder than the marvels of the modern world, yet they paved the way for these new systems.

A qanat is a traditional rainwater harvesting technique used in the Middle East.
Pafnutius, CC BY-SA 3.0 <https://creativecommons.org/licenses/by-sa/3.0>, via Wikimedia Commons: https://commons.wikimedia.org/wiki/File:QanatFiraun.JPG

A traditional rainwater harvesting technique used in the Middle East is qanats. These engineering masterpieces channel water from slopes underground into an agricultural area or to be used for drinking water. The same method has been applied in various North African and Asian regions. In Oman, the construction is called a falaj, and in North African countries that used the technology, it was referred to either as a *khettara* or a *foggara*. In Central Asia and Pakistan, the term *karez* is used. All these names described the same basic structure. Vertical holes are dug into a slope, which all feed into an underground horizontal channel. The channel leads to the bottom of the slope, where the water is collected in a storage tank or a dam or dispersed into irrigation channels. Modern water collection techniques have replaced the need for widespread use of qanats. However, new interest in the technology is peaking as a potential method for sustainable groundwater collection.

Stepwells are a popular way of collecting rainwater in India.
Jakub Hałun, CC BY-SA 4.0 <https://creativecommons.org/licenses/by-sa/4.0>, via Wikimedia Commons: https://commons.wikimedia.org/wiki/File:20191219_Panna_Meena_ka_Kund_step_well,_Amber,_Jaipur,_113 2_9646.jpg

In India and other areas in the subcontinent, the use of step-wells to collect rainwater was popular. Constructing a step-well is a commitment that requires many hands on deck. The structure takes up a large surface area to collect as much rainfall as possible. The spaces where step-wells were located were often communal, but recently, as technology and urbanization took place, they fell out of favor. A step-well is built in a square, with steps funneling the surface water to the narrow well in the center of the construction. Some of these old wells are elaborately decorated, showing how important they were to daily life for generations past.

Principles from these older structures can be applied to your personal water collection system, like using a large surface area to harvest water, manipulating gravity to your advantage, and constructing channels to transport your water underground on slopes. Some of these old principles have stood the test of time, but they can be combined with some modern innovations. The ancients were not only masters at rainwater collection, but they were also knowledgeable on multiple techniques of dew harvesting. In Israel, archeologists have discovered low circular walls built around vegetation to trap condensing water. South Americans and Egyptians used a similar method: pile up stones so that water vapor could cool in between the gaps under which a container was placed to catch the falling droplets.

It may be time to renew and improve some of these dew and rain harvesting techniques as multiple factors converge, making it a necessity in the future. With pollution and climate change, water will become scarcer, so individual families may need to find ways to collect rain. Furthermore, rainwater collection may create a more sustainable future as stormwater runoff is collected instead of letting the water flow, carrying pollution and contaminants. Moreover, surface runoff from water can cause soil erosion, which impacts food security. As financial instability increases all over the world, collecting rain and dew may also be a way to save money.

Underground storage tanks for water have proved to be a welcomed innovation due to the space that they save. In urban environments, space is a commodity that not many have access to. This problem has been solved by moving water storage underground so that more surface area can be available in cramped cities. Vertical rain gardens and green roofs have also gained popularity. Green roofs, which are also called living roofs, are gardens that are specially constructed on the top of a roof to grow flowers and produce. The soil acts as a sponge from which water can be collected. Moreover, green roofs can be energy-saving because they provide additional insulation in the winter. Vertical rain gardens are stacked plant setups placed near gutters to channel rainwater to produce food in limited spaces.

When it comes to dew harvesting, one of the central innovations in this field is the use of grooved surfaces. When grooved surfaces are used on a slope, the depth of the groove causes more condensation to occur, allowing for the collection of bigger droplets. This makes dew collection far more efficient and less labor-intensive. Fog collectors have also been introduced into some mountainous areas to collect low-lying clouds. The fog collector uses a mesh material made from synthetic fibers like nylon to catch droplets out of the atmosphere. These droplets are then channeled into plastic storage containers. Fog catchers only work in specific environments where it's moist and fog is common, but it is an easy and efficient way to collect moisture straight from the air.

Safety Considerations

Water collected from rain and dew is not immediately drinkable. Microbes and contaminants in the water could cause illness for humans and animals. You must be mindful of what you are going to use your rainwater for, like flushing toilets, washing, irrigation, or consumption. If you are consuming the water you collect, you have to be extra careful to make sure that it is drinkable.

Cleaning your water takes two steps, namely, filtration and disinfection. Filtering your water takes out all the bigger particles that may be harmful, like metals and dirt, while disinfection kills all the harmful microorganisms. There are different filtration methods that you can use. The filtration technique you use will determine how fine the particles that can go through your filter are. Cloth and sand can be used for filtration. Beach or river sand can be placed in a tube through which you allow your water to flow, or you can pour your water through some fabric. These filters are crude and should not be used for drinking or cooking water. You can purchase some great filters that attach to faucets and directly filter the water that flows from them.

To disinfect your water, you can either boil it or use chemicals. Boiling is an ancient, tried, and proven method that kills many microbes because they cannot survive in the harsh environment of those extreme temperatures. You should boil your water long enough to ensure that any harmful organisms are thoroughly taken care of. Using chemicals to disinfect your water is a little more complex because you must get the dosage right. Bleach and chlorine are two commonly used cleaning chemicals. However, they can be harmful for consumption if too much is used, so you must make sure you know the exact ratios you should mix. The taste and smell of chlorine can be removed from water by using specialized charcoal or carbon filters.

UV light is another way to disinfect your water. This method can be expensive and requires you to know how long your water must be exposed to the light and how much water one bulb can disinfect. This method is gaining popularity as UV disinfecting becomes more embraced socially. Distillation is another effective way to clean your water. Distillation is the process of boiling liquid until it evaporates, then recooling it to become liquid again. Distillation is expensive because you need a lot of energy to boil water to the point of evaporating it. It can also be used as a way to separate salt from water.

UV light can be used to disinfect your water.
Coleopter, CC BY-SA 4.0 <https://creativecommons.org/licenses/by-sa/4.0>, via Wikimedia Commons: https://commons.wikimedia.org/wiki/File:Uv_lamp.jpg

Before you go through the filtration and disinfection process, you must consider the equipment you use to collect rain and dew. Make sure that you keep your catchment areas clean and that you replace your filters regularly according to the manufacturers' guidelines. If you are using roofs as water catchment spaces, make sure that you clean them properly before rainfall or big storms because a lot of dirt and germs can gather on them when exposed to the elements. Lastly, remember to keep your clean water and the liquid you have just collected separately. Your tanks and storage barrels should also be made from materials that do not contaminate the water and should be cleaned regularly. If your tank is empty, run the first water collected in it out because it is likely to contain dirt that has piled up at the bottom.

Chapter 4: Rainwater and Dew Storage

Since collecting rainwater and dew are easy and cheap options for obtaining water, many people gravitate toward these methods. However, gathering the liquid is just the first part. Safely storing water and ensuring that it is potable could mean the difference between life and death. There are hundreds of waterborne diseases with excruciating symptoms that stem from consuming untreated, stagnant water or as a result of weak safety protocols. Your well-being is in your hands because you are solely responsible for finding out how to store water appropriately. With a basic understanding of the considerations you need to make when storing water, you can take the necessary steps to keep clean water over long periods.

Safely storing water is paramount to the success of your endeavor to survive independently.
https://pixabay.com/photos/metal-container-technology-8332370/

The containers you use, as well as your disinfecting and filtration techniques, play a role in storing a safe, drinkable supply of water. Your maintenance routine and the care you take in keeping your water contaminant-free will hugely impact your health. In an outdoor survival situation or on an off-grid homestead, safely storing water is paramount to the success of your endeavor to survive independently. By learning the details of what makes water safe to drink and what steps you can take to maintain your storage setup, you can thrive with an abundance of potable water.

Long Term Storage Containers

Unless you are living in a tropical region where it constantly rains, the wet season will pass by quickly. This means there is a limited window for you to gather as much water as possible. The way you store your water will impact how usable it is in the dry seasons. There is no expiration date on water, so you can store it as long as you need to. However, water is a carrier for all kinds of microbes, parasites, and contaminants. The containers you use to store your water are the first line of defense against disease-spreading microorganisms, so you'll need to put a lot of thought and research into the containers you use and not store them recklessly by pouring water into any plastic bottle you find lying around.

When you think about how rainwater is collected, it is easy to see how it can become dangerous for humans to consume. For example, if your tank is connected to gutters from your roof, bird excrement can be washed into your water supply, and no one would drink a glass of water with fecal matter floating around in it. You must think about the array of foreign particles that could wash into your water supply. Although collecting rainwater is convenient, it is not pure by any means. Rain can gather all kinds of dangerous pathogens from the surfaces that it touches.

The first stop when considering long-term water storage is your containers. The tanks and barrels that you keep water in must be food-grade. All plastics are not created equal. There are many different ways to manufacture plastics that contribute to the quality of the material. Food-grade plastic is of higher quality than containers not intended to hold water. If you look at the plastic containers that you use, there should be a number in a triangle. This number is called the resin identification code, or recycling number, and it tells you what kind of plastic your container is made from. Many food-grade plastics will have the code HDPE or

LLDPE, but this way is not the most reliable when determining if containers are safe for storage. The same kind of plastic that is used for water can also be applied in other instances, like for fuel, so be careful because it is doubtful that you want gas-flavored water. Safe RIC codes that you can store water in also include PETE and PET.

The FDA determines which plastics are safe for water and food to be stored in. Plastic does tend to leak into your water, so it is not 100 percent safe. However, some plastics are safer than others. Glass is probably the safest material to store water in. Still, it is impractical because there is no glass container big enough to carry huge supplies of water, and the containers are easy to break, so an extra level of care must be taken. However, a small glass container may be beneficial if you are outdoors in a wilderness survival situation, but there is still the risk of breaking the fragile material.

How well your containers are sealed also affects the contaminants that can enter your storage vessels. Your containers should be filled to the top so that there is no air in them. Moreover, your containers should be clearly labeled with what the water can be used for and the date it was stored. You could use some water for plants and flushing toilets, but it may not be appropriate for consumption. To avoid mixing up the uses for the different water you have stored, keeping them separated and clearly marking them is essential.

The water you store should be shaken or agitated at least every six months if not used. This may be enough time for the new rainy season to come back, but it isn't advisable to drink that water anymore if it is not. Drinking the water you have been using without incident throughout the year can be tempting, but all it takes is one slip-up for disaster to strike. You should be as careful with your water as you would be with a fragile newborn baby.

Some people have used canning jars for drinking and cooking water, which is also a brilliant solution for fighting microbes and reducing contaminants. These jars are easy to store and can be packed in boxes until needed. Since glass is the best material for storing water, tightly sealed jars are ideal. Furthermore, using canning jars also allows you to transport your water easily. Make sure that your jars are clean before you pour any water into them. Another benefit of using jars is that you can differentiate water meant for other uses and water meant for consumption because all your drinking water will be preserved in these sealed containers instead of in large plastic tanks.

Keeping Stored Water Free from External Contaminants

Getting the right containers is only one part of the battle against illnesses that come from water. Once your water is in your chosen containers, you still have to make sure that no external contaminants infiltrate your tanks, barrels, or jars. An air-tight seal and a container filled to the top are essential. Your water should also be kept in a cool, dry place that is between 50- and 70 degrees Fahrenheit.

Just like you must clean the surfaces from which your water is collected, you should also sterilize your containers. First, wash your vessels with warm water and soap. After you have rinsed all the soapy water out of the container, it is time for disinfection. Mix one teaspoon of unscented chlorine bleach to about four cups of water. Your bleach should range from 5 percent to 9 percent sodium hypochlorite. Shake the container well and let the mixture rest for about thirty seconds. After that, you should rinse the container thoroughly and let it air dry. This will eliminate any parasites or pathogenic bacteria that may have gathered inside the vessel.

If you store water for long-term use, it is likely that your water will be kept in bigger containers and then decanted into smaller ones. Any utensil you use to transfer your water from one vessel to the next should also be kept clean and sterilized. If you dip a dirty bucket into a water tank, the hundreds of liters in that tank are now also contaminated. Therefore, it is also advisable to never let your hands come into contact with your water supply because numerous pathogens are living on your skin. Humans can be a walking disease factory, so be mindful of your body when you interact with your water.

Your environmentalist leanings may prompt you to use recycled containers to store your water. This is fine as long as no toxins were previously stored in the vessel and the material is food-grade. However, you also need to ensure that there are no cracks or holes in your recycled containers. A damaged container can easily become a superhighway for disease. Inspect your containers before using them to ensure that they are in peak condition, especially if they are going to be used for long-term storage. This is also why you must seal your containers properly. Another helpful tip is to use some storage containers with narrow necks so you can pour the water out without touching any of the contents.

Some basic hygiene protocols are crucial when working with water, like washing your hands before handling any of your storage containers. Keep the environment where you store your water clean as well. It is better to elevate your storage to avoid getting contaminated with dirt or animal waste. Hygiene is key because you are fighting invisible enemies like parasites, protozoa, and pathogenic bacteria.

Some people use a pond to store their water.
https://unsplash.com/photos/a-pond-surrounded-by-trees-in-the-middle-of-a-forest-7CdNNvijw9E?utm_content=creditShareLink&utm_medium=referral&utm_source=unsplash

Some people may use a pond or a dam to store their water, but this method is more dangerous because your water is exposed to the elements. It is not advisable to drink water that has been stored openly in a dam. There may be other uses for this water, like watering crops or flushing toilets. If you have water that is being stored in the open like this, you must be extra careful to prevent it from mixing with your potable water. Water in dams tends to be cloudy, so it is the perfect breeding environment for microbes and living organisms like mosquitoes and mollusks, which also call stagnant ponds or dams home. Their waste could be dangerous for humans, not to mention that mosquitoes in some regions carry malaria.

Dangers of Water Stagnation and How to Prevent It

If you've ever watched a survival video or television program, you'll notice that when people are looking for water, they never drink from a stagnant source and opt to seek out flowing water like a river or a waterfall. Stagnant water is alive, but not in a good way. Insect larvae, parasites, and other microbes make their home in stagnant water. This is concerning because if your water is being stored in barrels, tanks, or other containers, there is a good chance that it is not flowing. Therefore, there are safety steps you must take to maintain your well-being when consuming the rainwater you have stored.

Stagnant water is dangerous for human consumption.
Bibiire1, CC BY-SA 4.0 <https://creativecommons.org/licenses/by-sa/4.0>, via Wikimedia Commons: https://commons.wikimedia.org/wiki/File:A_stagnant_water_at_Ilepo_Araromi.jpg

Stagnant water is categorized as the most dangerous class of water for human consumption. Water is classed into four categories, namely, clean water, grey water, slightly contaminated water, and black water. Stagnant water outside falls under the class of black water because the pathogens that it breeds are so dangerous to humans. Some of the deadliest diseases and parasites with horrific health costs, like salmonella and E. coli, are found in stagnant water. Viruses like hepatitis E and rotavirus also thrive in stagnant water. The mold that is produced in this black water can be dangerous. The combination of bacteria, viruses, parasites, and mold makes stagnant water something that you do not want anywhere near your property, especially close to your drinking water.

It is common for people to pump stagnant water into their plants, but experts recommend that it should be contained and expelled completely from a property. Mounted trucks can be called in to get rid of this water. Willingly introducing stagnant water onto your property is illogical. Store your water in sealed containers to prevent the laundry list of issues related to stagnant water.

Sealed bottles with narrow necks are your best option to prevent the contamination of water that isn't flowing. Even in this ideal situation, you must swap out your water every six months to a year. The tight cap will prevent unwanted visitors, and the vessels should be kept out of direct sunlight. Suppose your water is being stored outside in a dam you have constructed. In that case, you must take the necessary action to get your water flowing. You'll notice that algae starts to form if water is still for a long enough time. This is a bad sign that your water is not potable and can be dangerous if handled. There are portable pumps available that can be used to circulate the water in your storage dam. You will have to factor in the cost of electricity or fuel needed to power these pumps. The bigger your water reserves are, the more powerful the pump you need will be. Circulating the water supply for a few hours a day should be enough to prevent algae buildup, but it does not combat other microorganisms that are not as easy to see.

If you are not drinking water stored in an outside dam, you may be tempted to let the algae develop. However, there are many dangers associated with algal growth. It can clog up your pipes if you have irrigation systems and rust metal components on your homestead; it can make your animals sick and even kill some plants. Preventing algae from forming is a better option than looking for ways to treat it afterward because some of the algae eradication chemicals also contaminate your water.

Big storage tanks sometimes promote algal growth. Keep it out of the sun in a cool and dry place to stop it from growing in your tanks. You also need to flush your tanks, especially after the first rain, to get all the access debris that piles up in the bottom out. Algae will not grow in a properly sealed tank because you are depriving it of the air it needs to grow and spread. Shallow and still water is where algae thrive, so if you construct a dam, it takes extra effort and resources to dig it deep.

Maintenance Routine to Keep Water Potable

Like any machinery you run, your rainwater storage and distribution must be maintained. Everything may seem to be working well, but all it takes is one overlooked detail for your system to spiral out of control. The consequences of not keeping up with maintenance on water gathering and storage can be horrific. Illness and diseases can creep in if you don't keep on top of your repairs.

The first check that you should make is that your containers are not damaged and that they seal properly. Contaminants slip through the smallest cracks, so you must constantly check for leaks. Replace tanks or barrels leaking so your storage systems run at peak conditions. Some plastic containers have expiration dates on them, so if you are using water bottles for storage, you must check that your vessels are still up to date.

Cleanliness is the foundation of keeping water potable. Your runoff surfaces should be kept tidy, especially before it rains. Furthermore, big water tanks gather dirt and debris at the bottom, so you need to flush out these containers occasionally so they don't become attractive to pathogens. Any pipes, gutters, and fittings attached to your tank should also be spotless.

Even when you keep water in food-grade sealed bottles, plastic will leak into the water after a while, so rotating your water every six months is essential if you are keeping an emergency supply. Ensure your storage room is also disinfected with bleach, and your water bottles are kept off the floor. Filtration is a must, but your filters should also be changed regularly according to manufacturing guidelines. Filtering removes various parasites and debris from your water. Only drink clear water that has been filtered and disinfected so that you do not get any viral infections. Your containers must be kept clean, so remember to disinfect your vessels after each use.

Filtration and disinfection are aspects which must be taken seriously and which cannot be neglected. There are many disinfectants on the market specifically for drinking water. Chlorine and iodine tablets are available to clean your water, but remember that these methods should only be used on clear water and not murky water because some microbes can hide within dirt particles to avoid the disinfecting effects of these chemicals.

Keep all your clean water for drinking and cooking away from water meant for other uses. The water you use for your toilet may not be as clean as the water you consume. Labeling is an essential part of your maintenance routine, so you should replace any damaged labels and mark new containers you introduce into your water storage system. Transferring your water from one container to another regularly also helps prevent the water from tasting bad. Your hands and clothing should be clean when handling your drinking water, so institute cleaning protocols before you enter the area where you store your water.

Chapter 5: The Art of Water Purification

After collecting water, you may be tempted to drink or use it right away. However, rainwater is contaminated and contains germs and chemicals that can be harmful to you and your family. Purifying the water will eliminate all toxins and render the water safe for consumption. There are many purification methods that you can use; some are primitive, while others are more advanced.

This chapter covers the inherent risks of untreated water and different purification techniques with step-by-step instructions for each one.

Purifying the water will eliminate all toxins.
https://pixabay.com/photos/water-drops-falling-droplets-20044/

Risks of Untreated Water

Untreated water is water that hasn't been purified or filtered and contains sediments, chemicals, and microorganisms like parasites, viruses, and bacteria. Consuming this water can cause various health conditions and, in severe cases, death.

Health issues caused by untreated water:
- Headaches
- Fatigue
- Fever
- Polio
- Typhoid
- Hepatitis A
- Fluorosis
- Dysentery
- Diarrhea
- Kidney and liver damage
- Cholera
- Gastrointestinal problems
- Neurological problems
- Stomach aches
- Lead poisoning
- Dehydration
- Amoebiasis
- Arsenicosis
- Nausea
- Salmonella
- Trachoma (eye infection).
- Poliovirus
- Cryptosporidiosis
- Cancer

Drinking untreated water is particularly harmful for people with weak immune systems, the elderly, children, and infants. They are more likely to get sick and may even die from the chemicals and microorganisms.

Protect your family's health by purifying the water with any of the following techniques.

Boiling Water

Boiling is one of the oldest and simplest ways to purify water. Older generations have been relying on this method for centuries. Even with the invention of filters and mineral water, some still prefer to use it. You can use this technique anywhere since it is fast and doesn't require any equipment.

Boiling is one of the simplest ways to purify water.
https://unsplash.com/photos/person-pouring-water-on-white-ceramic-mug-A4Gy_rEdsdA?utm_content=creditShareLink&utm_medium=referral&utm_source=unsplash

Advantages of Boiling Water

- It is one of the quickest and most convenient methods. You can boil water in a kettle, microwave, stove, solar cooker, gas stove, gas grill, fire pit, wood stove, or survival stove.
- It is cost-effective.

- It improves the water's taste by removing undesirable odors and flavors.
- It kills viruses and bacteria and removes solids.

Disadvantages of Boiling Water
- In some cases, it changes the taste, giving it a strange flavor.
- It requires heat and energy, which can be inconvenient in regions where energy is expensive and heat is unavailable.
- It is time-consuming, especially if you are boiling large amounts of water.
- Boiling water doesn't eliminate many harmful substances like lead, chlorine, microorganisms, heavy metals, chemicals, hormones, fertilizers, pesticides, and microplastics.

Instructions:
1. Put the water in a big stainless steel pot.
2. Put it on any source of heat available.
3. Leave it for three minutes to boil.
4. Then let it cool down.
5. Store it in a clean and sanitized container and seal it tightly.

Clay Vessel Filtration

Clay vessel filtration is one of the world's oldest and most traditional water purification methods. Before using advanced methods like reverse osmosis and UV radiation, people treated their water in clay pots. Clay works by trapping mud, contaminants, and impurities, leaving the water safe. Although it is a primitive method, there are still people using it to this day.

Advantages of Clay Vessel Filtration
- Eliminates impurities, bacteria, and protozoa in water
- Improves the taste
- Enriches the water with healthy minerals like iron, magnesium, and calcium
- Environment-friendly
- Cost-effective
- Accessible and easy to use

Disadvantages of Clay Vessel Filtration
- Doesn't remove viruses from water
- The water is prone to recontamination
- Requires regular cleaning

Instructions:
1. Pour the contaminated water into a clay pot
2. Cover it with its plastic lid
3. Leave the water to filter
4. You can drink it from the clay container

Desalination

Desalination involves removing salts and minerals from seawater or any other type of saltwater. It makes water with high levels of saline safe for consumption. Although it is more common for commercial use, like on boats, in resorts, hotels, and residences, it can be suitable for individual use. This method works in areas that only have access to saltwater.

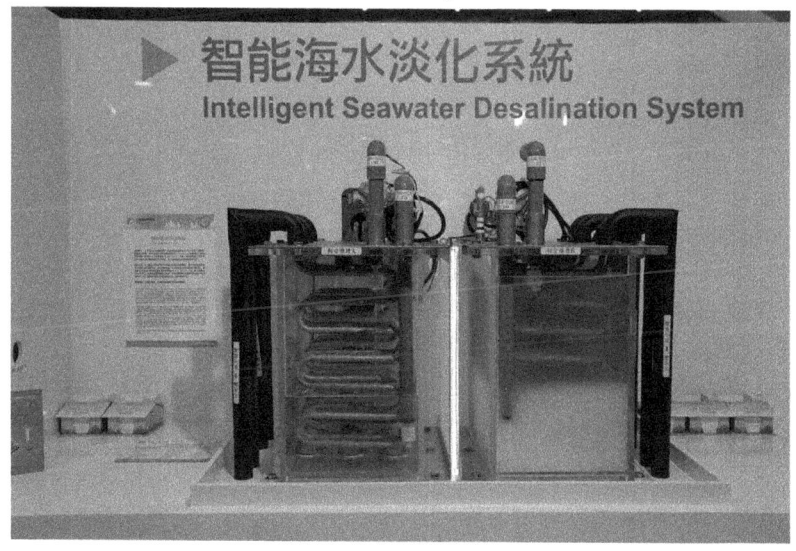

Desalination removes salt and minerals from saltwater.
Devan Hsu, CC BY-SA 2.0 <https://creativecommons.org/licenses/by-sa/2.0>, via Wikimedia Commons: https://commons.wikimedia.org/wiki/File:ITRI_Intelligent_Seawater_Desalination_System_20170603.jpg

Advantages of Desalination
- Provides clean and fresh water
- Removes salt from water

- Eliminates toxic minerals and compounds
- Separates microorganisms from the water

Disadvantages of Desalination
- Requires a lot of electricity (but not for individual use)
- High-cost

Instructions:
1. Get a large bowl for the salt water and a container for the condensed water. Make sure the bowl is much bigger than the container.
2. Fill the large bowl with salt water and put the small container in the bowl. It should be floating on it, and you must cover it with a paper bag.
3. Place them under direct sunlight, like outside or near a window.
4. Put a small rock on top of the paper bag to push the condensed and treated water into the cup.
5. Leave them for four hours.
6. You should find fresh water in the container. This is the treated water that is safe for use.

Distillation

Distillation is a water purification technique that involves the use of steam and condensation by boiling the water and converting the steam to liquid. The process eliminates any microbes in the water but doesn't remove metals, solids, and some pollutants. It is ideal for purifying large amounts of water but requires heat or solar energy.

Advantages of Distillation
- Perfect for people who live off-grid since it doesn't require the use of electricity
- Produces clean and high-quality water
- Removes salt from the water, making it safe for consumption
- One of the safest water purification methods
- It doesn't use filters or chemicals
- Removes bacteria, fluoride, and heavy metals

Disadvantages of Distillation
- Takes longer than other methods
- It isn't cost-effective
- Quality may vary depending on the equipment you use

Instructions:
1. Put the water in a large pot and put an empty small pot inside it.
2. Leave the water to boil on a stove or under the sun.
3. Leave it on medium heat.
4. Place the lid upside down over the large pot to allow the condensed water to trickle down into the small pot.
5. Then, place ice on the flipped lid. The difference in temperature on both sides of the lids will increase the speed of the condensation process.
6. When the ice melts, add more.
7. This will take some time, depending on the amount of water you use.
8. You will know it's done when the large pot is empty and the small one is filled with the treated water.

Iodine

This method is a little controversial since you are trying to remove chemicals from the water and not add more. However, life off-grid and emergency situations won't always give you a lot of options. So, only use this method when you can't use any of the other techniques. The best chemicals for treating water are iodine and chlorine.

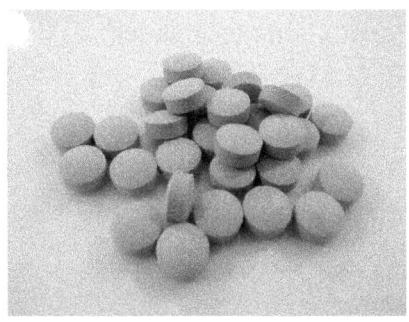

Iodine tablets can eliminate viruses and bacteria from water.
Mx. Granger, CC0, via Wikimedia Commons:
https://commons.wikimedia.org/wiki/File:Iodine_pills.jpg

Iodine is a red chemical that is sold as a liquid or in tablet form. It eliminates viruses and bacteria from water but leaves an unpleasant taste. If used in high dosage, iodine can be fatal. So be extremely careful when using it, and it should only be a last resort.

Advantages of Iodine
- It is convenient and cost-effective
- It kills protozoa, viruses, and bacteria
- Lightweight
- Easy to use

Disadvantages of Iodine
- Strong aftertaste
- Not safe for pregnant women
- Fatal in high dosages

Instructions:
1. Add five drops of 2% tincture of iodine to one liter of water and 10 drops if the water is colored or cloudy.
2. Stir the water, then leave it for 30 minutes to an hour before using it.

Ion Exchange

This technique works on removing arsenic, nitrates, radium, and barium from the water through ion exchange. The great thing about it is that it treats water without affecting its taste. This technique is very powerful since it removes heavy metals that many other methods fail to eliminate.

Advantages of Ion Exchange
- You can use it with other techniques
- It can be used on a large scale
- Easy to use
- It is safer than many other water treatment techniques
- Removes bad odors and tastes from the water
- Easy to install
- Easy to maintain

Disadvantages of Ion Exchange
- The stored water has a short shelf life

- Expensive
- Uses chemicals like salt
- Can't eliminate bacteria, viruses, pyrogens, or particles

Types of Ion Exchange
- **Water Softening:** It removes minerals that harden the water, like magnesium and calcium, and replaces them with sodium ions to soften the water.
- **Deionization:** It removes anions, cations, and all ions, resulting in purified and uncontaminated water.

Instructions:
1. Install an ion exchange water filtration system.
2. Once installed, the water will be safe for use, and you can consume it right away.

Reverse Osmosis

Reverse osmosis (RO) is a water purification technique that separates dissolved solutes from water. It eliminates heavy metals, microplastics, PFAS, VOCs, arsenic, chlorine, sediment, salt, fluoride, herbicides and pesticides, unwanted molecules, ions, and other contaminants. It is one of the most popular methods of purifying drinking water. People have been using this method for decades, and it is considered one of the most significant inventions in modern history.

Advantages of Reverse Osmosis
- Removes chemicals, germs, bacteria, viruses, and other biological contaminants
- Small and easy to use
- Doesn't require electricity or energy usage
- Improves water waste
- Fully automotive
- Removes impurities like fluoride, chlorine, and lead
- It does not use any chemicals to purify the water
- It is extremely popular in commercial use because it is adaptable and can cater to several needs
- Suitable for private and public use
- Produces high-quality water

Disadvantages of Reverse Osmosis
- Produces a lot of waste
- It is costly and requires regular maintenance since it can be clogged
- Eliminates all types of minerals, including the good ones your body needs
- Using it regularly can sometimes lead to cardiovascular issues, weakness, muscle tension, and fatigue
- Doesn't remove microorganisms like bacteria and viruses
- Doesn't disinfect the water
- Requires more energy than other methods
- Produces waste

Instructions:
1. Install the reverse osmosis system in your home or RV under the sink or outside. Or you can hire someone to do it for you.
2. This is an automatic method that will purify the water by itself.

Shungite for Water Purification

Shungite water purification isn't as popular or commonly used as the other methods in this chapter, but it is a very interesting one. Shungite is a rare mineral that you can only find in Russia. Some scientists believe it came from a meteor that crashed into Earth centuries ago. It contains high levels of fullerene carbon, a rare carbon structure that can absorb contaminants from the water.

Advantages of Shungite Water Purification
- It has antiviral and antibacterial properties
- Destroys viruses and bacteria from water
- Kills microbes in water
- Removes dangerous pathogens that are considered health hazards
- Purifies water
- Protects against EMF exposure

Disadvantages of Shungite Water Purification
- It can release chemicals and heavy metals into the water

- This method is a little risky, so caution is advised

Instructions:
1. Get one or more Shungite stones.
2. Wash them thoroughly for two minutes under water.
3. Then, wash them again five times.
4. Soak the stones in the water and leave them for five days.
5. Change the water every day.
6. Remove the Shungite from the water, put it in a container, and then cover it with the water you want to purify.
7. Leave it for 24 hours to seep.
8. Then remove the stones and drink the water.

Solar Purification

This technique treats water using UV radiation. Solar energy removes contaminants from water, reduces microorganisms, and prevents them from reproducing, making the water safe for consumption. It is one of the simplest methods you can use. However, it won't work in cold regions or places without direct sunlight access.

Solar purification isn't as powerful as other techniques, so use it only when you don't have other options.

Advantages of Solar Purification
- Removes microbes from water
- Improves water quality
- Environmentally friendly
- Cost-effective

Disadvantages of Solar Purification
- Doesn't work in cool regions or cloudy weather
- Has no impact on extremely contaminated water
- Doesn't remove chemicals from water
- Only works with small amounts of water

Instructions:
1. Fill a plastic bottle with water.
2. Shake it to activate the oxygen.
3. Place it horizontally under direct sunlight.

4. Leave it for an hour.

UV Radiation

In this method, you use UV bulbs or lamps that emit UV light that can eliminate some microorganisms. However, this method isn't strong enough and can't remove heavy metals or impurities.

Advantages of UV Radiation
- Kills protozoa, bacteria, viruses, and microbes
- Environment-friendly
- Easy to install and use
- Requires little maintenance
- Doesn't use much electricity
- Doesn't waste water
- Water is instantly purified, so you can consume it right away
- Cost-effective

Disadvantages of UV Radiation
- It isn't effective against all microorganisms
- You will need to filter the water before using this method
- Requires electricity, which isn't always available for people living off-grid
- Doesn't enhance water's taste or odor
- Doesn't remove metals or solids

Instructions:
1. Put the water in a UV water purification machine and let it do its job.
2. The machine exposes the water to UV light, which kills microorganisms' DNA and prevents them from reproducing, making the water safe for use.

Water Chlorination

Chlorine is the second chemical you can use in treating water. Water chlorination is one of the oldest purification techniques. It is ideal to use during emergencies since it is simple, fast, and effective. It involves adding a mild bleach with 5% chlorine to the water to remove toxins and microorganisms.

Advantages of Water Chlorination
- Deactivates microbes and eliminates harmful microorganisms
- Disinfects the water, making it safe for consumption
- It is easy to use and brings fast results
- It is cost-effective and doesn't require any equipment

Disadvantages of Water Chlorination
- Leaves a strange odor in the water
- Constant exposure to chlorine-treated water dries and weakens the hair
- Causes skin irritation
- Chlorine changes the water's taste

Instructions:
1. You will need chlorine bleach to disinfect the water. Choose plain liquid laundry bleach like Purex or Clorox.
2. Check the ingredients and avoid using bleach with scents and additives.
3. Determine the amount you will need in advance. Use it moderately to avoid any issues.
4. Use four drops of chlorine for every 10 mils of water, or 8 drops if the water is very contaminated or cloudy.
5. Add the chlorine to the water and mix them well together.
6. Leave the chlorine water for 6 to 12 hours. Don't use the water before this time.
7. Test the water for chlorine using a digital meter.
8. Make sure the water is clean and chlorine-free before using.

Water Purifier

Water purifiers eliminate water impurities, minerals, and biological contaminants. Although they require electricity, there are many options on the market that you can use off-grid, like a pump water purifier. This is one of the most effective methods since it removes more contaminants than other techniques. However, it still doesn't remove chlorine and lead.

Advantages of Water Purifiers
- Environment-friendly
- Cost-effective
- Improves odor and taste
- Improves health
- Provides safe and clean water
- Kills microorganisms
- Easy to use

Disadvantages of Water Purifiers
- Water purifiers require maintenance
- They don't remove pesticides

Instructions:
1. Install the pump water purifier using the instructions on the package.
2. Fill it with water.
3. Pump the handle up and down for a couple of minutes.

Traditional vs. Advanced Water Purification Methods

Now that you have familiarized yourself with different purification methods, you are probably wondering which one is right for you. The answer to this question depends on many factors, such as the area you live in and if you live in a cabin, house, or RV. So, consider all these factors before making a choice.

However, the advanced ones take the lead when comparing traditional to advanced methods. Thanks to technology, there are newer, safer, and better methods to purify water. For instance, a traditional method like boiling water isn't always safe. Although your grandmother would argue that her family has used it for hundreds of years, recent discoveries have shown its limitations.

Things have changed in the last few years, and there are new types of viruses, bacteria, and heavy metals that traditional methods struggle against.

Advanced methods are better in every way than their traditional counterparts; although they may be costly and some can be less convenient, they are the safer option. You use water in every part of your life, so you shouldn't take any risks.

There is no denying that living off-grid has its challenges, and during an emergency, safety can be a luxury you can't afford. In this situation, you will use what is available to you rather than what is safer. However, if you can be prepared in advance, always choose safety.

Water purification isn't optional. It is a necessary process to remove chemicals and microorganisms from your water. Make sure to treat the water right after you collect it using the appropriate method. Stick to the instructions and don't change or ignore any of the steps, especially with chemical doses.

What would you do if you lived in a cold environment where all the water sources were frozen? The next chapter provides tips and skills to turn snow and ice into water without risking your health.

Chapter 6: Snow and Ice: Melting Tips, Myths and Misconceptions

In this chapter, you will learn all about carefully sourcing water from ice and snow, as it is often the only lifeline in colder areas. This chapter will equip you with the knowledge and skills to make informed decisions when sourcing water from ice. You will learn all about the energy conservation techniques for melting snow and about the debunking of related myths and misconceptions. You will also be learning about the proper way of treating ice while avoiding the risk of hypothermia. Let's dive right into it:

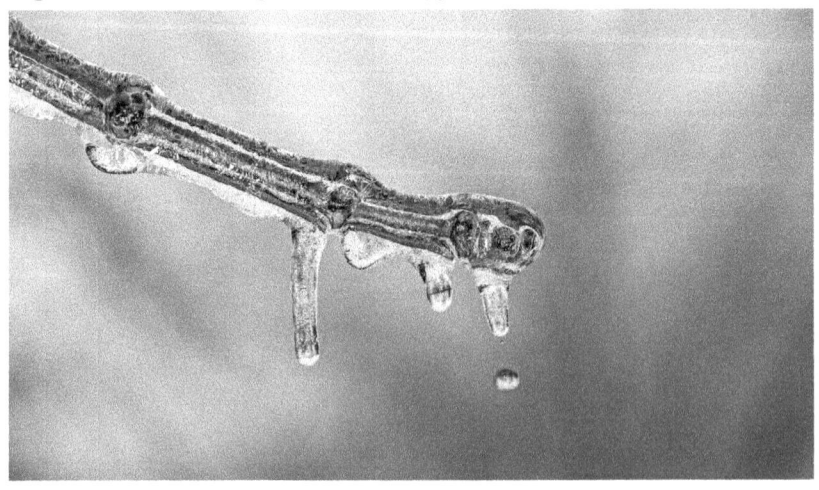

Turning snow into drinking water to keep yourself hydrated is an essential skill to master if you live in cold environments.

https://pixabay.com/photos/ice-melt-frost-melting-frozen-570500/

Efficient and Energy-Conserving Techniques for Melting Ice and Snow

Turning snow into drinking water to keep yourself hydrated is an essential skill to master in cold environments. It is also an essential skill that hikers and mountaineers must also learn. You can eat or drink snow by filtering it properly. You can consume a small amount of ice or snow for survival. However, ingesting large quantities can be extremely dangerous.

Consuming snow can lower your body temperature as your body requires a significant amount of energy to convert ice and snow into water, and the energy used causes your body temperature to change. You must keep your body temperature high, especially if you're hiking long distances. This is why it is critical to use efficient ways to convert ice to water. Let's take a look at some of them below:

Use a Stove

Many outside venturers travel in ice-cold environments with their modern stoves that require liquid fuels such as Coleman fuel, propane gas, or unleaded petrol to melt ice and snow. Don't fill your pot to the top with snow. It's a great insulator and may cause your pan to burn rather than melt the ice. The most efficient way to do it is by collecting the snow from outside your campsite and then melting a small amount of it in your pan on a gentle flame (Ensure that the snow you're collecting is not yellow). Add small amounts of snow to the pot until you have enough water. The snow mainly comprises air, so it takes a significant amount to produce a liter of water. To save fuel, you can also cover the pot or pan with a lid. You can also heat it over a fire if you have trees in your surroundings. The water from snow is not safe just because it is frozen. Bacteria and pathogens may become dormant when frozen but will be reactivated when the ice melts. Hence, it is preferred that you boil your water before consumption. Practice the same discretion with standing water.

Heat It Over a Fire

You can also melt your snow over a fire if you do not have a stove or a pan. You may hang the ice and snow over the fire and let the melted water drip into a container. You may wrap the ice in a cloth, a shirt, or any other porous item that can be hung with the support of sticks or even ski poles. You can also heat a hard sheet of ice or crusty snow over a fire without using any vessel. You may cut the sheet of ice into smaller pieces and put it over the stick to melt it.

Use the Sunlight to Your Advantage

It is alright if you do not have access to fuel or fire; it's still possible to melt ice without fire. You can also use the sun to melt your ice. When the dark rocks come in contact with sunlight, it becomes easier to melt ice or snow that comes in contact with them. You can find water even in temperatures as low as -10 degrees Celsius. Listen out for flowing water on cliff faces or south-facing slopes. Once you find your water, you can collect it in a container. It is important to find it and collect it before dark, or it may freeze again once the sun goes down. You can keep your water bottles in your sleeping bag to prevent them from freezing again.

Use Body Heat

If you carry a water bottle, keep it in contact with your body or inside your coat to keep the water from freezing. Take some snow and put it in your bottle whenever you sip your water. Your body heat (and the water in your bottle) will melt it quickly and replenish your water supply. Ensure you keep a wide-mouthed bottle so it is easier to put snow in it. This is one of the easiest methods to get water and does not require you to make a fire. It also works in emergencies or bad weather conditions.

Use a Zip-Loc

Another way to melt the snow under the sun is by putting it over a garbage bag called a drum liner. You place your snow in a Ziploc and then place it over the garbage bag. The bright heat from the sun will melt the snow inside the Ziploc and give you clean drinking water. This method works best if you are staying in areas with a bright sun. However, if you are not carrying Ziplocs, you can just place the snow on the drum liner and wait for it to melt. You can then put the water in a container. This process is difficult, and collecting the melted water this way may become a hassle.

Melt Ice on a Tilted Rock

This is another easy way to melt ice. You can melt snow with some rocks and a fire. Create a tilted table-like surface to hold your ice so that you can heat from below the rocks with fire. Then, channel that melted snow into a container below the rocks. You must use clean snow so you don't need to filter the water. It is important to use rocks that are not porous. Otherwise, the water will soak into the rocks, and you will lose it. You should also avoid taking the rocks from near the river, as they are more likely to become highly saturated with water. Suppose you are in areas like Coastal Alaska or British Columbia. In that case, you will be better off melting the sea ice that is found at a distance from the coast rather than the recent ice.

Eating Snow Directly

This should be the last resort, as eating a lot of snow can dangerously lower your body temperature, which increases the risk of hypothermia. However, very cold ice can burn your mouth and lips.

Melting Tips

It requires a lot of energy (i.e., fuel) to melt snow, as snow melts slower than ice, and fuel will probably be scarce. This is why poor melting methods can lead to excessive use of fuel. Here are a few melting tips that can help you conserve energy:

Protect Your Stove from the Wind

You must find a sheltered place for your stove as the wind can increase the amount of fuel that it takes to boil a liter of water. If you have a lot of snow around you, then try cutting snow blocks to create a windbreak for your stove. Although it isn't perfect, it will still protect your stove from wind. You may also use bushes or piled rocks to protect your stove from the wind. You can then use your own body for extra protection, too.

Heat with Water in Your Pot

Melting snow in a dry pot can be a slow process. You should always try to keep a little bit of water in your pot for efficient energy transfer to the snow. First, add an inch of water to the pot and then put in the snow. The snow will melt, and the water level will rise. Then, pour out the water from the pot while keeping the same amount, one inch of water at the bottom of the container.

Do Not Heat the Water

If you aim to just melt snow and not heat water, you must not waste energy heating water. Keep removing the liquid water from the pot as it melts while adding snow to the pot. You can add snow and stir the pot before you take out the water, as it helps recover the energy lost while heating water.

Use a Lid

Putting a lid over your pan or patting while melting the snow is a great way to conserve energy. It is quicker as well compared to melting the snow without a lid. If you put a lid on your pot, you trap the hot air and steam inside the pot, which fastens the melting process and significantly enhances the melting efficiency. It also reduces the risk of spillage while stirring the spot.

Establish Redundancies

If you are going on an outdoor adventure, ensure you have a working stove, fuel, and a pan, as these will make it easier for you to source water. If you do not have this equipment, there are other passive ways to melt snow, as discussed above.

Debunking Myths and Misconceptions

Although snow can be a wonderfully convenient water source, it is important to debunk some myths and misconceptions related to it. It is important to clarify these misconceptions for your and your loved ones' safety. Being aware of these issues helps you make informed decisions. Let's jump right into it:

Myth #1: Eating Snow Can Hydrate You

Reality: Eating snow can be great for extreme survival conditions, but you should not rely on it fully. People tend to misunderstand that snow is made just out of water, so it can be a great source of hydration. However, it is not as simple as it seems. Although snow is made of water, many good properties are lost in the water-freezing process, which is why eating snow can still result in dehydration instead of preventing it, as the body uses more heat and energy to melt the cold snow. This can lead to a reduction in your body's fluid level while making you more dehydrated than before. It also does not contain any essential nutrients and minerals you will need for proper hydration. However, melted snow is a great water source in emergencies and if you're outdoors with no access to streams.

Myth #2: Snow Is Purified Upon Melting

Reality: Snow does not automatically purify upon melting. Although it does remove some impurities, it does not eliminate all contaminants. Boiling the melted snow or using other water purification methods is the best way to remove harmful chemicals, pollutants, and microorganisms.

Myth #3: Snow and Ice Are Not Equivalent

Reality: While it is true that snow and ice can be melted to yield water, their processes may be different. Snow is composed of air, so it takes more time for the snow to melt than dense ice. This is why you should be patient and take care to melt ice and snow in the correct way.

Myth #4: All Methods of Melting Snow Are Equally Safe

Reality: All methods of melting snow may seem equally safe, but that is not true. If you try to melt snow in a plastic container or a contaminated container over an open flame, harmful pathogens and contaminants may be introduced during heating. Using a camping stove and a clean container is highly recommended.

Myth #4: Yellow Snow Is Safe to be Melted and Consumed

Reality: Yellow snow can be quite dangerous to consume as it may contain impurities and contaminants such as urine. It would be better for you to avoid drinking melting yellow snow as the presence of contaminants and harmful substances can be hazardous to your health.

Myth #5: Snow Is a Limitless Source of Clean Water

Reality: Snow cannot be considered an infinite water resource as, depending on your location, it may not even be available all year round. It should only be used in emergencies or when no clean, flowing water is available.

You need to educate yourself on safely consuming melted snow as a water source. Even in emergencies, take appropriate measures to ensure that the snow or ice you consume is clean. It is best to use the proper water purification method, such as boiling or water purification tablets, to treat the melted snow.

Potential Hazards of Consuming Snow

Sourcing water from melted snow and ice requires caution. It is essential for you to carefully melt the snow and purify it to avoid any health risks. Let's discuss the potential hazards of consuming melted ice water without proper treatment.

Hypothermia

You must beware of Hypothermia. Hypothermia refers to the state of your body when its core temperature drops below the normal temperature required for healthy metabolism and body functions, which is 95 degrees Fahrenheit (35C). All temperatures below 95F are considered hypothermic; however, it may vary:

- Mild = 90-95F
- Medium = 82-90F
- Serious = 68-82F
- Acute = 68F and below

The risk of hypothermia increases if you ingest snow directly, as it significantly reduces your body temperature. As snow's temperature is below freezing point, it can make your body expel extra energy to bring the temperature of snow down to normal body temperature. This leads the body's core temperature to enter hypothermic levels, easily worsening your bad situation. Hence, you should always try to melt the snow into water before drinking it.

You can make your water by putting the snow into a container or a bottle of water to make it melt and be good enough to consume. Please only use white, fluffy, and clean snow that does not seem dirty or contaminated. Purifying the melted water by boiling it before consumption is also a great idea.

May Contain Contaminants

It can be extremely dangerous for your health if you consume melted snow water without proper purification, as it may be contaminated. The water may contain pathogens like bacteria, viruses, and other chemicals. These harmful substances can get released into the snow from industrial and agricultural waste, animal waste, and air pollution. You should also know that the snow may also contain heavy metals, including mercury, lead, and arsenic. All these harmful substances can be released into the water when the snow melts, making it unhealthy to drink. The presence of these unhealthy substances in your water can be really bad for your health. This can cause many health issues as well, including neurological issues, gastrointestinal diseases, kidney problems, and even cancer. There may also be organic compounds in your water that can cause respiratory illnesses.

High levels of chlorine in your water can also be unsafe. If consumed in large quantities, it can lead to nausea and vomiting and may even cause liver damage. You can use a proper filtration system to remove bacteria, viruses, and other contaminants to make the water safe to drink. Boiling water even for one minute can make it safe for consumption.

May Lead to Dehydration

As discussed above in one of the myths of consuming melted snow water, drinking melted snow can lead to dehydration. As snow consists of a small amount of water, it can lead to quick loss of fluids and eventual dehydration. If you are going to consume snow, you must do it in moderation, as your body temperature may drop quickly if larger amounts are consumed at once.

Poor Sterilization

Properly sterilize the melted snow water for your safety's sake. As mentioned above, the snow may contain many contaminants, so you should try to purify the water properly through boiling or purification tablets. Poor sterilization may lead to the causation of waterborne diseases.

Melt snow carefully to avoid potential risks from consuming contaminated melted snow water. You must take sufficient precautions to ensure the water is safe for consumption. If possible, try boiling, filtering, or adequately melting the snow before drinking it. Taking these steps will significantly reduce the risks to your health.

Chapter 7: Conserving Water in Scarce Regions

This chapter focuses on strategies and practices to optimize water use in areas where resources are limited. After detailing the challenges unique to arid or drought-prone regions and the dire implications of water wastage, the chapter outlines actionable steps to minimize water usage, from daily activities to agriculture and industry. You'll learn about traditional water-saving practices, innovative solutions, emerging water-conserving technologies, and community-driven efforts and policies that have made a tangible difference in water-scarce regions worldwide in short and long-term scarcity situations.

Conserving water is essential.
https://pixabay.com/photos/to-protect-hands-ecology-protection-150596/

Water Scarcity and the Importance of Water Conservation

Due to a combination of extreme population growth, inefficient water management, pollution, and climate change, water scarcity has become a pressing issue in several areas of the world. Even if you don't live in a region of global water scarcity, contributing to conserving this highly valuable resource could mean a vast difference in the future. From hygiene issues to economic losses through unsustainable agriculture to the dire consequences for the Earth's ecosystem, water scarcity enormously impacts many areas of life. This problem is particularly notable in arid regions due to high evaporation rates and limited precipitation. Because of this, over 50% of the world's wetlands have already disappeared over the last century. Wetlands are the most sustainable habitats on the planet as they support the survival of many species. They're also the primary cultivation place for staple crops like rice and support a variety of ecosystems that benefit people, including flood control, water filtration, and storm protection.

In areas of water scarcity, all-natural landscapes suffer. Freshwater lakes are getting smaller and becoming salty due to pollution and because too much water (compared to the overall volume) is used for irrigation. Seas are retracting, leaving behind polluted land, creating food shortages and a rise in poor hygiene-related and waterborne diseases, and lowering the local population's life expectancy. Moreover, without adequate clean water resources, people can't get enough for consumption and agriculture, which leads to economic decline.

However, water conservation isn't only needed in drought-prone areas. Finding sustainable solutions to this issue is vital for improving water availability and conserving this resource. It also helps promote natural ecosystems, preserve habitats, and prevent saltwater intrusion into groundwater in coastal regions. Moreover, it can help save money for individuals and their communities and for entire countries by reducing the energy consumption needed for water transportation and treatment.

Practical Solutions for Water Conservation

Building Dams and Reservoirs

Establishing reservoirs like artificial lakes after river dams is an efficient way to collect water during high precipitation periods and conserve it for dry periods. However, they also have another purpose. They can be used as water supplies for urban areas, flood control, hydroelectricity creation, and more. Unfortunately, dams aren't the most ecological solution as they lead to downstream river channel erosion and other negative impact on the local ecosystem. The water in these reservoirs warms up more quickly, which means it evaporates rapidly and becomes unsuitable for species adapted to cold river water. Moreover, the sediment trapped in the damp creates a need for further filtration, depending on what the conserved water is intended for.

Reservoirs can help you collect and conserve water.
Naksh, CC0, via Wikimedia Commons: https://commons.wikimedia.org/wiki/File:Water-Reservoir-51552-pixahive.jpg

Desalination

Advanced desalination (removing salt and other compounds from seawater) technologies have become increasingly popular in coastal regions. Salt can be extracted from brackish and seawater through distillation or reverse osmosis, rendering it suitable for agricultural uses, like irrigation, human consumption, and industrial purposes. As saltwater sources are readily available in coastal regions, desalination represents a sustainable resolution to freshwater scarcity in these areas, reducing the dependence on fast-disappearing resources and helping meet the needs of coastal populations. Benefits notwithstanding, desalination has drawbacks, including brine disposal issues, which can affect the ecosystem and high energy consumption. Solar-powered desalination is a more cost-effective variant, as it uses less energy and can be applied in smaller regions.

Aqueducts

Installing aqueducts is another way to transfer water from one place to another. However, they only work in small areas because drawing water from larger areas could lead to drought in these areas in the future. These dry areas lead to the loss of ecosystems and significant air pollution in the surrounding region. Moreover, transferring from larger areas would mean crossing a vast distance, which raises the cost of the process.

Rainwater Harvesting

For centuries, people have been collecting rainwater for agriculture, recharge, and human use. In areas where rainfall is scarce, harvesting and storing it can be a fantastic solution for improving water availability during dry spells. The traditional method of rainwater harvesting involves collecting water from rooftops and similar surfaces that catch and lead large quantities of water toward the ground. This inexpensive approach can be used in off-grid situations as it can be combined with water purification techniques to make water safe for human consumption.

Beyond offering a secure water supply during droughts, the method also diminishes the reliance on other water resources. If an entire community starts implementing rainwater harvesting, they'll have a significant impact on improving local ecosystems in nearby wells, lakes, and rivers while contributing to long-term water management. Groundwater recharge, for example, enables water to move from the earth's surface to the deeper layer of the ground. Harvesting rainwater is a viable solution to create more groundwater and reduce water scarcity. Advanced rainwater harvesting methods go a step further. These also eliminate soil erosion, flooding, and other problems created by sudden stormwater runoff in urban environments and help use this excess water for building long-term storage instead.

Water-Smart Agriculture

Water-smart agriculture refers to targeted farming practices for boosting or retaining crop productivity while maintaining optical water usage efficiency. This approach involves both traditional agricultural methods and modern water conservation efforts like planting drought-resistant crops, drip irrigation, or precision agriculture.

Drip irrigation delivers water into the ground, directly to the crop's roots, maximizing water efficiency and minimizing losses due to evaporation. By enabling farmers to apply precise water delivery control and conserve resources, drip irrigation can lower water consumption by up to 50% while optimizing crop yields.

Xeriscaping is an agricultural method involving using crop fields requiring minimal water use and maintenance that uses water. These landscapes are created through a combination of drought-resistant plants, optimized mulching applications, and efficient irrigation systems to lessen water consumption. Beyond preserving water, xeriscaping also diminishes the need for pesticides and fertilizers, resulting in healthier and safer crops for human and animal consumption.

Adopting these and other water-smart agricultural practices can be an excellent way for farms to lower production costs, improve water retention in soils, contribute to sustainable water resource management, and reduce evaporation rates. These agroforestry practices have an overwhelmingly positive impact on ecosystem resilience and biodiversity preservation.

Pollution Control

Uncontrolled water pollution can lead to even greater water scarcity, especially in drought-prone areas. To prevent freshwater resources from becoming unsafe to drink and having to look for other solutions for an increasing number of people, sewage systems should be improved, and water quality should be regularly monitored and monitored.

Wastewater Treatment

Implementing efficient wastewater treatment solutions is another way to reuse and reclaim contaminated water. These techniques remove all the contaminants from water, producing a fresh supply of reusable water. In most cases, this water is used for industrial or agricultural purposes. However, with the right technique, it can be purified to a level that becomes safe for human consumption. Moreover, modern water purifying techniques create a natural gas byproduct that can produce energy. By investing in this cutting-edge technology, people can contribute to a sustainable infrastructure for continually reusing resources and securing their water needs.

A form of water waste treatment is greywater recycling, which entails collecting used water from different sources in a household, including sinks, washing machines, and showers. This water is typically used for irrigation, groundwater restoration, industrial purposes, municipal water supply, and toilets, all of which reduce the dependence on freshwater supplies while reducing wastewater dump.

The best part of greywater recycling is that it can be implemented for both large-scale and small-scale users. For instance, in coastal areas, wastewater from a company's or municipality's waste system can be used for groundwater supplementation, limiting saltwater intrusion. Small-scale use is seen in single households, for example, where the plumbing is designed so that greywater from sinks is used to flush the toilet. For the most part, greywater is filtered until it becomes non-potable, which is unsafe for conception. However, if this treated water is sent through additional purification and aquifers, it can become potable (drinking) water.

Water Conservation

Water conservation entails using water more economically. For individual households, this means tailoring user behavior to use less water and implementing water-saving technology. The former can include washing dishes through a dishwasher instead of hand washing them, running the dishwasher and the washing machine only when these are full, so they require fewer washes which use less water than small frequent loads, taking short showers instead of baths, turning off the water while cleaning your teeth, or fixing leaky fixtures. Additional water conservation requests could include opting for produce with a reduced water footprint, like eggs, instead of high water footprint variants like beef. Likewise, growing drought-tolerant plants in your garden in dry climates, irrigating only as needed and preferably in the morning before the sun gets high and evaporation begins, can save surprisingly high volumes of water. Water-saving technology includes installing water-conserving shower heads and toilets and more efficient washers.

Since agriculture uses far more water than households, an even greater effort for water conservation can be made here. Besides the drip irrigation system, another viable solution could be growing only crops native to the area and only in places with enough precipitation to sustain them or using no-till methods, which rely on soil coverage application to reduce evaporation.

Smart water metering is an innovative water conservation technique that can be implemented in various fields. Industries, municipalities, and even individual households can monitor their water usage through real-time water consumption measurement. This enables users to pinpoint areas where they can curb or modify their water usage and management. By identifying excess water usage and implementing the necessary measures to reduce this and become more responsible, users can reduce their consumption by up to 20%.

Encouraging Water Conservation by Educating Others

While individual users can significantly contribute to water conservation, they can do much more to secure water supplies in the long run by educating those around them. By collaborating with others in your community, you can radically reform water consumption in the entire community. Educating others involves raising awareness about the importance of water conservation. They can also be encouraged to get involved in the process and provided with practical tips on saving water on their properties. Individuals should also aim to stay current with the innovative water conservation technology available in their area and how these can help resolve water scarcity issues. This fosters a sense of responsibility for their own behavior.

Educating community members about the impact of their water consumption habits and encouraging economical water use practices can be done in several ways, including media and social media outlets, school programs, etc. Community-wide awareness campaigns are another method for teaching individuals of all backgrounds about the importance of appreciating the value of limited freshwater resources.

The positive effects of collaborative efforts for this purpose have been proven time and time again, and not just on a community level. However, it all starts there. When communities raise an issue, other organizations and governments will notice it and start putting in the effort to exchange resources, expertise, and knowledge that brings the necessary changes to resolve the issue.

Past experience shows that governments that created and implemented enforceable legislation and policies to incentivize water-efficient practices have made a vast difference. Some of the best ways this was achieved was by fostering more responsible water management in industrial sectors and individual households and promoting the use of highly efficient technologies.

Not only can communities collaborate in water preservation efforts, but so can countries. Freshwater resources are often shared by at least two countries. Thus, it makes sense that they cooperate in managing this very finite resource. Promoting collaborative efforts like joint initiatives and water-sharing agreements could foster a greater sense of collective responsibility.

Taking Advantage of Artificial Intelligence

Artificial Intelligence (AI) has revolutionized many sectors, including water management. Utilizing AI to supplement water conservation efforts helps people to make more informed decisions on researching and developing new water-saving technologies and techniques. AI can help analyze water consumption data and optimize water management strategies based on predicted water demand, weather forecasts, insights on optimal irrigation schedules in agriculture, possible issues (like leaks, for example) in building infrastructure, and more.

By leveraging these tools, entities can make better decisions regarding conservation efforts, investment in improving infrastructure, and water allocation, addressing large-scale water scarcity challenges and securing sustainable water use.

Real-Life Examples from Around the World

Langtang Region, Nepal

As remarkable as the landscape in the Eastern Himalayan regions is, the rugged terrain and climate changes have caused many headaches for local farmers. The weather shifts from extreme droughts through the winter to prolonged monsoon seasons through the warmer months. It makes it harder and harder for the locals to grow crops and sustain their livelihood and the traditions that have depended on the area's natural resources for centuries. On top of that, the mountain glacier lakes are melting, threatening floods and the reduction of freshwater resources that currently sustain over a billion people. It could additionally destroy the habitat of numerous species already endangered by human actions.

To address the issues mentioned above in their area, communities in Nepal's Langtang region, along with the support of outside organizations, developed a project to help them better adapt to climate changes, including unpredictable precipitation patterns, water scarcity, and shifting seasons. These changes included installing drip irrigation systems (extended across multiple communities), along with water storage to preserve rainwater, establishing seed banks to supply farmers with more drought-tolerant species, and education programs to teach them how to implement techniques like crop rotation and reduce water footprint when raising farm animals.

Additionally, the efforts included planting shrubs and trees along slopes to prevent landslides and endangering freshwater resources. The villagers were given more efficient household appliances, and newly established weather stations gathered data to monitor subsequent climate change. All these efforts helped educate the community, making a vast difference in the villagers' lives and ensuring their livelihood and the livelihood of their families. The project did not only address the effects of climate change but also the needs of people and the local ecosystem. It raised people's awareness and gave them a sense of responsibility for preparing for the inevitable future effects of climate change. It also served as an example to other Himalayan communities facing the same challenges, which are now working on implementing similar changes.

Sahel Region, Africa

The Sahel region of Africa faces a similar problem, receiving rain only three months of the year. However, while this wasn't usual in the past, the periods of rain have become unpredictable due to accelerated climate changes. During the months without precipitation, the soil dries out completely, becoming so hard that it can't even penetrate the surface layers when the rain arrives. Instead of giving it much-needed hydration, the water simply slides off the surface and evaporates under the scorching sun. Consequently, local communities struggle to secure water for the better part of the year for agriculture and hygiene, cooking, and drinking. Fortunately, they received help from several organizations and learned how to conserve water during those short months of the rainy season so they could have water to use throughout the rest of the year. For example, they've learned to build pits (zai or half-moon pits) – essentially holes dug into the soil to trap rainwater and prevent it from sliding off the hardened surface. Stone binding barriers are another example of the effort to keep the water on the agricultural lands.

Trees around lakes are planted to stabilize the land and prevent it from being washed away during heavy rainfalls. Several communities are working on digging even larger water storage pits. These 10-foot-deep and 100-foot-wide reservoirs represent more targeted efforts as they will be established after a thorough evaluation of where water is collected during heavy rainfalls, calculated based on the terrain and the water's natural flow. These tanks will be able to hold the same amount of water that could fill an Olympic swimming pool, providing a substantial resource for families struggling to meet their freshwater needs throughout the year. Due to their size, the water in these reservoirs won't be affected by evaporation despite the extremely high temperatures and the unpredictable rainfall.

The benefits of the large rainwater pits in the Saharan regions will be short- and long-term. By putting water conservation techniques in place, the local communities can secure access to fresh water all year round, contributing to public health and safety and ensuring the productivity of the lands that sustain their livelihood. Moreover, the soil quality around the reservoirs improves as the rain gets trapped in the deeper layers. People can start growing crops in previously dried-out lands and obtain more nutritious food resources. This effort can be further bolstered by planting an outer ring of trees around the reservoirs. Trees also trap water and provide shelter for the crops, preventing the water used for their irrigation from evaporating. This shows that water conservation is not only beneficial for having a resource for a community to tap into but can also transform the entire environment and create new, more sustainable ecosystems.

Chapter 8: Surviving While on the Move

Detailing the unique challenges and strategies associated with maintaining hydration and sourcing water when continuously mobile, this chapter provides a better understanding of the preventive and reactive measures to stay hydrated while moving through diverse terrains and situations. From water rationing to purification to reading natural signs that can hint at nearby water sources, you find everything you need to ensure long-term water sustainability without guaranteed resupply.

The Importance of Staying Hydrated While Being Mobile

Whether you are forced to be on the move for migration, long-distance trekking, or fleeing emergencies, staying hydrated is vital, especially outdoors. Knowing how to avoid dehydration is key to enjoying outdoor activities safely and staying healthy in emergencies. Water is needed for many metabolic and physiological processes in the body, many of which are critical for survival. Depleting your body's water storage results in fatigue, loss of focus, heat exhaustion, and much more. For all these reasons, staying hydrated is critical for preparedness in extreme conditions or crises.

Staying hydrated is critical for preparedness in extreme conditions or crises.
https://pixabay.com/photos/faucet-fountain-water-1684902/

At the same time, the wilderness and other emergency situations make it far more challenging to maintain appropriate hydration levels. However, with proper preparation and by learning what precautions to take when facing extreme conditions and limited water resources, you can avoid dehydration and its consequences.

Tips for Staying Hydrated on Outdoor Adventures

Drink before You Start Moving

If you're going on an outdoor adventure, like hiking or a short, day-long camping trip, drink one or two cups of water, even if you aren't thirsty yet. When you start feeling thirsty, your body's water levels are already depleted enough for dehydration to begin. Drinking enough water before you go out will keep your body's water levels in an optimal range for longer. If you regularly hydrate before starting your day or activity, this will improve your fitness levels, enabling you to focus better while outdoors.

Suppose you're heading out in the morning. In that case, you can start hydration the evening before by ensuring you won't go to bed thirsty and avoiding caffeine and alcohol at nighttime. The latter two cause dehydration, so these are not great options to drink during camping or other outdoor adventures.

Make Your Portable Water Supply Easily Accessible

Make sure you carry ample water for your adventure. Besides this, you'll also need food, as this is a source of electrolytes and essential salts – which are just as crucial for preventing dehydration as drinking enough water. Sweating causes your body to lose electrolytes and leads to an imbalance in your hydration levels. For a short hike, a bottle of water and a few energy bars or other snacks or dried food will suffice, while a camping trip involving mealtimes requires food and water for each mealtime. If you're going on an extended adventure, count on meals and water for every day, plus a little more in case something comes up and your return is delayed. Electrolyte tablets are another option for these situations, and they can come in handy for fast rehydration in a crisis where food isn't available.

To ensure you'll be drinking and eating regularly during your adventure, make your resources easily accessible. While many prefer to carry water bottles, these make it harder to drink on the go. Bladders, however, can be easily used while on the move, ensuring you won't forget to use them. If you still prefer bottles, put them where you can easily reach them. For example, many hikers and campers prefer forward packs with forward-facing pockets instead of side pockets for water bottles because they make reaching for water easier. Using water-storing containers with built-in handles and attached carabiners is another alternative. You can clip them onto your belt or backpack and make them easily accessible while on the move. Likewise, keeping snacks in easily accessible pockets will make it more likely for you to reach them.

Drink Water Regularly

Instead of drinking when you feel thirsty, drink ½ -1 quart of water every hour while outdoors. Suppose the temperatures are high, and you're moving through demanding terrain. In that case, you'll likely need more because moving in warmer climates raises your body's water intake needs. If you're bored drinking water this frequently, you can alternate between pure water and electrolyte-filled fluids, like sports drinks. This will help you retain optimal hydration levels and make your outdoor adventure more enjoyable.

Stay Hydrated Afterwards, Too

Continue replenishing your water and electrolyte levels even after you return home, and don't wait until you get thirsty. Remember, thirst isn't a good indicator of your body's proper water needs, so it's always better to drink a little more rather than not drinking enough.

Signs of Dehydration

Dehydration should be treated as soon as you feel it because the longer you wait, the greater the consequences. Here are some signs of dehydration to look out for in yourself and those who are with you in your outdoor adventures or emergency situations:

- **Dry Mouth**: Before your body even registers thirst, you'll probably notice you have a dry mouth.
- **Thirst:** Feeling it means your body is telling you that you are dehydrated and should drink immediately because your water levels are about to become dangerously low.
- **Low Energy Levels:** You may assume you're tired because you're on the move a lot, but this can also be a sign of dehydration.
- **Muscle cramps, Headaches, and increased fatigue** are severe signs of dehydration. If you feel these symptoms, you must rest and rehydrate immediately.
- **Dizziness, Nausea, Stumbling, and Fumbling** are likewise serious symptoms, indicating that your body is struggling to provide enough fluids for normal metabolic and physiological function.
- **Dark Urine and Abdominal Pain:** Your kidneys suffer from a lack of hydration, making the urine concentrated and struggling to filter metabolites properly.
- **Weight Loss**: Even if you can't weigh yourself regularly, if you notice that your clothes fit looser after strenuous activity, it means you've lost water weight, and this water needs to be replenished through fluids and electrolytes.

What to Do in Case of Dehydration

If you notice the signs of dehydration, stop and find shade (if this is possible). Take a few moments to rest, especially if you have moderate signs like fatigue or muscle cramps. If you have more severe symptoms, you'll need to rest longer, and if the temperature is extremely warm, apply body cooling methods as well. Soaking a shirt or any other piece of clothing in water and placing it on the back of your neck, wrists, or head are the quickest ways to cool down. Meanwhile, drink water mixed with electrolyte tablets (if nothing else is available, table salt will do as well). Oral rehydration and salts help the body absorb water more effectively. Adding them to your first aid kit (a highly recommended part of your gear for wilderness adventures) won't take much space, and they can be a lifesaver.

A compact quick water purifier is valuable as it can make an enormous difference in your ability to rehydrate yourself or anyone else in case a fresh water supply isn't available. While you should always aim to have enough water, sometimes it can be difficult to judge how much water and electrolytes you need depending on the weather conditions, not to mention the possibility of emergencies in which you may need to use water for other purposes. Having a purifier handy will make rehydrating from alternative water resources easier without exposing you to pathogens and toxic substances.

Additional hydration tips for every outdoor situation:
- If you have plenty of freshwater sources on your route, drink about 33 ounces of water every time you stop to refill. This way, you'll remain hydrated while keeping your load light.
- Drink regardless of the weather. Unless you are forced to ration your supply, proper hydration is just as critical in cold weather as it is when temperatures are high. Not only does cold cause dehydration, but being dehydrated will make you feel like you're freezing. Use a thermos to keep yourself hydrated and prevent your water from getting chilled during the winter.
- Exposure to the sun will also aid dehydration, so wearing sunscreen and sun-protective clothing is essential for summer outdoor adventures. Even shade from your clothes and hat will help you regulate your body temperature, sweat less, and remain hydrated.
- It's a good idea to go on outdoor adventures in extremely hot climates when it's cooler. If you're on the move due to an emergency situation, try keeping to the shade when the temperatures are the highest and walking around in the morning or early evening when the sun is no longer at its peak.
- Rehydrate with at least 16 ounces of fluids (ideally mixed with electrolytes) overnights or when resting midday during the warmest hours.

What to Do if You Run Out of Water

If you notice that your water supplies have diminished or are about to run out, don't panic. Remaining calm will help you maintain your focus on finding alternative resources. Panicking is a waste of energy and time. It also increases your body's evaporation rate, making you lose liquids you can't afford to relinquish. As a first step, assess your situation and area. If you've just run out of water and it's not extremely hot outside, you'll likely be fine without water for a few hours. It may get uncomfortable, but if you've hydrated properly until then, it shouldn't be dangerous for your health. However, if you're further than that from the nearest water source or battling scorchingly high temperatures, you'll need to start searching for alternative solutions immediately.

Finding and Using Water in the Wilderness

Access to water is crucial for surviving in the wilderness. You'll need to find alternative sources when your supply becomes limited. Here are a few tricks on how to do that.

Look for Vegetation

Fruit can be a great water source - as long as you know which to look and reach for, as not all are safe for consumption. Familiarize yourself with the native flora in the area you'll be moving through during your outdoor adventure or mobile survival situation to learn which fruit and plants to look for as a water and electrolyte resource. Due to their high water and essential nutrient content, citrus fruit, grapes, berries, apples, and melons are the most recommended options - but there are many more you can reach for rehydration, including edible grass and moss variants. To determine whether a plant is suitable for this purpose, examine it for signs of water storage. For example, some plants store water in their nodes (where the leaves grow from. These are typically swollen when they're full of water), in the leaves (look for thick, moisture-laden foliage), or underground bulbs and tubers. The vegetation's color, shape, and size also play a role in determining its suitability for safe rehydration. For instance, brightly colored plants are often toxic, so you should avoid them. Moreover, some plant parts require preparation for either extracting water or rendering them usable for consumption.

Vegetation can lead you to a water source.
https://unsplash.com/photos/green-plants-and-trees-during-daytime-GwDOjh66Lc8?utm_content=creditShareLink&utm_medium=referral&utm_source=unsplash

Tapping into trees is another efficient strategy for finding drinking water in the wilderness. Trees with edible sap, like maple and birch, are great options as they contain larger quantities of readily drinkable fluids. They are also easier to tap, and extracting the sap takes only a few minutes. For example, other trees, like walnuts, are also viable options, but extracting sap from them is far more challenging.

In arid areas, any sign of vegetation indicates moisture. Looking for livestock or other animals is another option, as these could lead you to a nearby water source (but be cautious when approaching or following them).

Find Flowing Streams

If you have rivers or other flowing streams in the area, these can be an even more reliable source of drinking water. Because water travels at higher speeds, flowing water sources contain less bacteria than standing water reservoirs. However, you still must be cautious when deciding which stream to use. For example, if you find evidence of animal or human waste near the source, or the water isn't particularly clear (fresh, unpolluted flowing streams are always so clear you can see through them), you should probably stay away from it - unless you have no other option and have the ability to purify the water.

An additional tip: If you find a small stream that's dried up, tracing its path upwards to its source can help you find an additional stream you can use for hydration. If you can find current water sources, you can try digging for water in moderately moist areas near dried-out streambeds.

Collect Rain and Snow

Rainwater can be a priceless resource for drinking and hygiene in unexpected situations during outdoor adventures and emergencies. To collect rainwater, set out large containers like buckets or barrels to store it until needed. You can also gather rainwater on the top of your tarps (or if you don't have any containers, you can hang strategically placed waterproof materials instead for collecting precipitation). Although rainwater is relatively clean, it should be further purified to eliminate any bacteria or other pollutants it may contain. The same applies to snow, another alternative water source, especially in mountainous regions. Snow and ice can be used only, if not too gray or yellow, and should be melted before consumption, as the cold fluid can lead to dehydration.

How to Purify Water

If you don't have access to clean water in the wild, your next best option is to purify the available resources, making them safe to drink, cook, or prepare medicine if needed. Depending on your specific circumstances, you can opt for several purification techniques, including the ones listed below:

- **Boiling:** The simplest and often most effective way to purify water. You'll only need to strain it through a small hole sieve to remove bigger chunks of impurities and bring it to a rolling boil for several minutes to kill all the bacteria and other hazardous microorganisms.

- **Activated Charcoal Filters:** Moving the water through activated charcoal filters will bind the harmful organic compounds and remove unpleasant odors and tastes from the liquid.

- **Purification Tablets:** Given that these tablets can clean large amounts of water in a short period, they are an excellent addition to any outdoor survival gear. Place one tablet into half a gallon (or a little less if the water is particularly contaminated), wait 30 minutes for the tablet to neutralize the harmful compounds, and your water will be ready for consumption.

- **UV Purification:** UV purifiers emit bacteria-killing radiation that can eliminate these microorganisms from your water supply in minutes.

Making the Most of Your Water Supply

Carrying Enough Water with You

Carrying enough water is one of the most fundamental factors to consider when preparing for an outdoor adventure or emergency survival situation. First and foremost, you must assess how much water you'll need for the time frame you'll be out on the move. This depends on several aspects of your situation, including the required physical activity levels, climate, the difficulty of the terrain, altitude, whether you'll be traveling during the day or night, whether you can expect obstacles in finding reliable food and water sources, and many more. For example, a person with a moderate activity level hiking on a demanding terrain on a warm spring day will need 33 ounces of water for every 2 hours of activity.

Refill Whenever Possible

Filtration kits, gear, and bottles make it easy to replenish your water resources when you walk past or near a freshwater resource in the wilderness. Look for collapsible options, as these can be stored in your bag when empty and used when needed.

Before heading out, research possible water sources nearby, but in the wilderness and from human origins. For example, you might find houses or camping grounds with water outlets nearby or temporary abandoned dwellings with running water or stored water resources. Have a plan in case anything goes wrong in an emergency or survival situation. Knowing where to go for a refill makes it easier to remain calm and preserve your energy and fluids.

Rationing Your Supplies

Suppose you are still worried about running out of fresh drinking water or don't have time to devise a refilling plan (or to research the area for available resources). In that case, you may need to ration your supply until you develop an alternative solution.

At moderate activity levels and temperatures, drinking only when you're thirsty is one of the best ways to preserve water. Rationing your loss rather than your supply will help you stay hydrated, waste less, and worry less about your dwindling resources. Covering up in hot and cold water will lower the risk of dehydration, and so will remaining in the shade. Your supply will last longer with fewer losses due to the sun, wind, and other elements.

Another tip to minimize evaporation from your body is to keep your mouth closed whenever possible. Unlike breathing through the mouth, breathing through your nose results in a slower and smaller evaporation rate. If you struggle to keep your mouth closed, try pretending that you're sucking on a round candy. This will automatically maintain your mouth in a closed position. Sucking also fosters increased saliva production, which will prevent your mouth from feeling dry, making you crave water less often. However, avoid actually sucking on pebbles or buttons because these can be a choking hazard if you are severely dehydrated and unfocused.

Smoking also adds to dehydration through evaporation, so you should avoid it if possible. Coffee, teas, and other drinks with a diuretic effect increase the amount of urine produced by the kidneys, leading to significant water loss. The same applies to salty food and drinks because these draw water away from cells.

Another crucial tip is to start saving water in time. Don't wait until you are severely dehydrated to ration your supply! By implementing water-saving methods from the get-go, you can save more efficiently and avoid getting into potentially dangerous situations.

Chapter 9: Long-Term Water Storage and Safety

In survival situations, whether it's dealing with a natural disaster, a prolonged emergency, or an outdoor adventure, having a plan for keeping a stable water supply is key to your well-being. This chapter will cover the ins and outs of long-term water storage. You'll learn about the methods for doing so and the important things to consider when it comes to maintaining a reliable water source for an extended period. How do you make sure the water you store remains safe to drink for weeks, months, or even years? What factors affect how long your stored water stays good, and how do you reduce the risk of it getting contaminated or going bad?

From understanding water quality to choosing the right storage containers and managing environmental factors, this chapter will prepare you to face the challenges of keeping a sustainable water supply, blending scientific know-how with real-world advice. Remember that in the world of survival, thinking ahead for the long term is just as crucial as having enough water.

Keeping a stable water supply is key to your well-being.
https://pixabay.com/photos/storage-tanks-vats-metal-tanks-20959/

Water Preservation

In the previous chapters, you learned the difference between disinfection and sterilization when it comes to water treatment. The goal is to make raw water safe for consumption by either killing or removing harmful components. However, it's crucial to understand that even the methods outlined in this book, including commercially available ones, don't produce sterile water completely free of germs or viable germ stages.

Even if you were to achieve sterile water, it would quickly become recontaminated when transferred to a different container, which could have stray dried-up algae, protozoa, or spores from the air in it. The methods described previously aim to purify or reduce the germ count in raw water, making it safe for consumption. With improvised methods, you can realistically reduce the germ count by a factor ranging from a hundred to ten thousand. Even sterilization in a laboratory rarely eliminates germs entirely, usually resulting in a reduction by a factor of one million at most, even under ideal conditions.

Now, let's talk about recontamination. The drinkability of water treated with improvised methods is not a given and depends on the individual's health and conditions. After the sanitizing process, the remaining germs continue to multiply, and their growth follows an exponential pattern. The speed at which this happens depends on factors like nutrient levels, temperature, and the initial germ count.

While there's no immediate need to drink or discard the treated water after a few hours, it's essential to consider that the remaining viable germs will continue to multiply. However, the risk can be minimized by storing treated water in clear conditions, carbon-filtering it if possible, and keeping it in a shaded or cool place. This generally ensures it remains safe to drink for weeks without harmful pathogen levels. This is why the storage of water is a process not to be taken lightly.

One remaining concern is bacteria that can tolerate normal temperatures and have modest nutritional requirements, like Legionella. However, the risk of harmful levels is significantly reduced with proper storage conditions and limited nutrients. Below are some of the most effective ways to preserve water and limit its contamination:

1. Tyndallization

Tyndallization is a laboratory process used to disinfect sensitive solutions. It involves repeatedly heating the solution to boiling point and then holding it at body temperature (98.6°F or 37°C) for a period called *incubation*. Boiling eliminates all relevant living germs but doesn't remove nutrients from the water. The heat triggers inactive endospores to "germinate" upon cooling, transforming them from a resistant dormant stage into a less resilient active one. They are then killed when the water is boiled again. This cycle is repeated multiple times. In a modified form, this method is used, considering that laboratory solutions, like nutrient broths, have a higher risk of rapid recontamination than relatively nutrient-poor drinking water.

Water heated to boiling can be consumed immediately after cooling, but it quickly recontaminates as nutrients from destroyed cells become available. Briefly reheating the water over the next two or three days reduces the number of inactive and heat-resistant stages to very low levels. After Tyndallization, the water can be stored and consumed at room temperature for several weeks.

2. Repeated SODIS

After the correct application of Solar Disinfection (SODIS) for a sufficient period, water remains free from harmful germs for several days. UV light, the primary agent in SODIS, can even kill stable endospores and oocysts of cryptosporidia and giardia. However, if the raw water is still cloudy or if germs are hidden, recontamination with bacteria is a possibility.

Exposing water stored in suitable containers to direct sunlight for an hour or two at midday kills any new or remaining germs or endospores. After repeating this process two or three times, the water can be transported without refrigeration for several days or weeks, depending on its nutritional content. This method is especially useful for prearranged desert trips, where clear PET bottles filled with reservoir water can be stored directly in the sun. However, prolonged exposure to UV radiation can age the plastic material, so these stores should not be left for more than a year. The water inside remains drinkable.

3. Chilling and Shading

Water treated with an "unreliable" germ count reduction method is likely to still contain algae, leading to the growth of fresh biomass in the presence of sunlight, nutrients, and dissolved gases. This results in a musty, putrid flavor and a slimy, foul-smelling layer (biofilm) forming inside the container. While this water is not immediately unfit for consumption, it becomes a breeding ground for any remaining viable, infectious germs over time, especially at temperatures above 59°F (15°C).

To counteract algal and bacterial growth, storing the water in a cool and dark place is crucial. This can be done by covering it with wet sand, which chills the container through evaporation, or by wrapping it in a rescue blanket and placing it in cold, raw water. Depending on the initial contamination level, water treated with "unreliable" methods can be stored for around two to three days.

4. Silver Ions and Other Chemical Substances

Similar to disinfectants, there are chemicals available for preserving drinking water. However, these agents are not considered reliable disinfectants. While disinfectants with oxidizing substances can kill and destroy pathogens, they cannot prevent the regrowth of any potentially remaining germs permanently. The reactive properties of oxidizing substances mean they do not remain stable for long once activated.

Water Storage Techniques and Ideas

1. Emergency Water Pouches

In today's world, convenience is everything; you can find practically anything prepacked and ready to go. Even drinking water comes in individual ready-to-drink portions—small pouches or packets holding about 4.22 ounces (125 ml) of water. However, these can be relatively pricey, ranging from 3 - 10 US dollars per liter. While marketed for their long shelf life and touted as emergency reserves for your car or boat, there's a more cost-effective and equally stable alternative: regular small water bottles, usually 330 or 500 milliliters (11.2 or 16.9 fl. oz.), found in any supermarket worldwide.

The water in these bottles is actually sterile, evidenced by its extended shelf life. You can safely ignore any use-by date printed on them—water safe to store for months remains just as safe to drink after decades. The only changes you might notice are a slight plastic taste and a slow reduction in quantity due to diffusion.

In contrast, "normal" flavored or sugary drinks have a shelf life. Over time, decomposition sets in, rendering the contents unfit for consumption. Carbonated water loses its fizz gradually, as the bottles aren't airtight, but other than a taste change, the quality remains relatively unchanged.

Here's the scoop on emergency water pouches: they're essentially tap water sealed in shrink-wrapped pouches. The good news is you can easily create your own at home. Just vacuum-seal ice cubes in a plastic bag, let them defrost, and "autoclave" them for half an hour in a pressure cooker. Consider storing fruit or iced tea mix powder in sealed jam jars or pouches if you like the idea of having flavored or sugary liquids on hand for emergencies. Add a bit to your water when needed, and you're good to go. It's a budget-friendly and practical way to ensure you have emergency hydration options without breaking the bank.

2. Transport and Storage Containers

In emergencies, ensuring a reliable and substantial water supply is critical. While water is often a transient necessity, certain situations demand preparedness. Stationary storage tanks are a valuable option for stand-alone systems in remote camps, disaster areas, or buildings without water supply connections. These large tanks, often intended for temporary use but capable of permanent installation, can be filled naturally using rainwater-harvesting systems. Due to extended standing periods, the water they hold must be treated as raw water rather than fresh rainwater, allowing for the storage of significant amounts over months or even years.

In the past, rural areas and off-grid regions stored water in cisterns, but today, plastic tanks are more prevalent. Both have pros and cons. Building a stone or concrete cistern presents challenges, while plastic tanks, such as IBC tanks with a capacity of around one cubic meter (250 gallons), are more susceptible to frost damage.

- **Cisterns and Ponds:** Ideal for holding large volumes of water, but even initially, clean water must be treated as raw due to its open nature.
- **IBC Tanks:** A practical choice for stand-alone systems. Sealable and stackable but vulnerable to freezing and bursting in low temperatures.
- **Buried Plastic Tanks**: A mid-sized solution that can be buried below the frost line. Less prone to bursting if it freezes, they can be filled with rainwater or covered with a roof to prevent contamination.

3. Mobile Medium-Size Storage Containers

Mobile storage containers, commonly known as *canisters*, become indispensable for those traversing arid regions by car or cart. With a carrying capacity ranging from five to twenty-five liters (one to six gallons), these containers are versatile and valuable for storing locally collected or harvested water in water-scarce regions.

Two primary options for transporting water are rigid plastic or metal canisters and folding water carriers. It's advisable to steer clear of collapsible water containers popular among campers because their thin walls are prone to breakage along folding lines during prolonged use.

- **Rigid Water Containers**

Rigid containers, while sturdy and easy to carry, come with a caveat—they occupy the same space, whether full or empty. Stackable and relatively robust, they often feature an integral tap that may flip open or snap off. Some prefer new "jerrycans," typically used for fuel, to avoid residual fuel contamination risk. Ensure food-safe canisters with quality lid seals are chosen to prevent leaks and maintain water quality.

Plastic containers have drawbacks, as they can absorb tastes and odors from adjacent canisters. Clear labeling and segregation (fuel/raw water/drinking water) are essential, and water containers should be stored away from fuel and lubricants.

- **Water Carriers**

While more susceptible to damage than canisters, water carriers have a distinct advantage; they only take up as much space as the water they hold. When empty, they can be conveniently rolled up and transported. Two common types include the Swiss Army water carrier (20 liters or 5.28 gallons), made from robust rubber with a solid spout, and water bags with a large filler cap, crafted from thinner material and available in various sizes.

- Swiss Army water carriers, while reasonably priced, can heat up in the sun, affecting water taste.
- Water bags made from synthetic material, like those from Ortlieb (a German manufacturer of outdoor equipment), are lighter but more expensive and prone to damage.

Bottles and Drinking Vessels

- **Transparent PET Bottles**: Lightweight, affordable, and robust, they're excellent for simple drinking and storage. However, they're heat-sensitive and unsuitable for boiling water.
- **Wide-mouthed polycarbonate (PC) Bottles:** Ideal for hot water and boiling without deformation.
- **Metal Bottles:** Allow for controlled water consumption and can be heated in a campfire. Suitable for serious physical separation and treatment methods.

Metal bottles allow for controlled water consumption.
https://unsplash.com/photos/green-hydro-flask-tumbler-on-wood-slab-ktpymCAIvGc?utm_content=creditShareLink&utm_medium=referral&utm_source=unsplash

Choosing the right bottle depends on water availability and expected treatment methods. Transparent bottles are preferred in water-deficient areas, while metal bottles are suitable for physical water treatment methods. Always consider the specific demands of your journey and the type of water treatment you plan to apply.

Improvised Transportation and Storage Methods

For backpackers with limited space and transport options, especially those on long journeys, water bags become a handy choice for collecting raw water or carrying emergency drinking water. But there's a catch - water bags have their limits, especially if unexpected rainfall offers a chance to store more significant amounts. Here's where some bushcraft experience can come in handy: you craft containers from bark, reed, bamboo, or wood. Remember, though, that natural materials can swell when wet and crack when dry. That's where thin, lightweight plastic sheets come to the rescue, serving as a reliable travel companion for extreme and long-distance travelers, facilitating the safe transportation of water.

1. **Crafting Storage Ponds with Simple Tools**

For those seeking an improvised storage solution for a temporary camp, a rescue blanket or similar sheet can be transformed into a small "pond." Start digging a shallow pit, ensuring the ground is free of sticks and sharp stones. Compact the soil thoroughly to prevent it from giving way under the water pressure. The pit's dimensions should allow the sheet to extend beyond the edges, giving you storage of fifty to a hundred liters (fifteen to twenty-five gallons) of water. These improvised vessels, created with natural materials, can be swiftly fashioned with lids, combined into transportable batteries, and even repurposed for cooking.

2. **Emergency Containers for On-the-Go**

If space in your backpack is scarce, choosing a lightweight water bag is not feasible. In such cases, you can resort to purchasing cheap wine bladders, often found in wine boxes. These are made from thin plastic but are prone to damage. Alternatively, large freezer bags and condoms, often part of emergency kits, can be used. However, these containers risk bursting even with minimal impact. Make a simple cloth bag from a T-shirt or unzipped trouser leg to stabilize them. Tie one end of the cloth in a tight knot, place the empty waterproof material inside, and fold the top seam to protect it. Fill the improvised carrier with water, ensuring no air pockets remain, and secure it with a knot or string. Lift the outer bag, tie it to the inner bag, and voilà - a stable, improvised water carrier ready for transportation. It's advisable to carry it outside the backpack or in your hands to avoid accidental spills and make the most of your resourceful water-carrying solution.

Additional Tips for Water Storage

Water storage is a critical aspect of preparedness, and understanding the nuances of storage practices in diverse environments ensures a reliable and safe water supply. Factors such as temperature, light exposure, and the threat of pests play pivotal roles in determining the efficacy of storage methods. Below is a comprehensive guide outlining best practices for water storage in different environments:

1. Temperature Considerations
   - Cold Environment
     - **Insulated Containers**: In colder climates, insulation is key. Use double-walled or insulated containers to prevent freezing.
     - **Underground Storage:** Burying water containers underground helps maintain a more stable temperature, preventing freezing.
   - Hot Environments
     - **Dark Containers**: Opt for dark-colored or opaque containers to minimize light exposure and algae growth.
     - **Shaded Storage**: Keep water containers in shaded areas to reduce temperature fluctuations and evaporation.
2. Light Exposure
   - UV Exposure
     - **Dark Containers:** UV rays can degrade water quality. Choose dark-colored or opaque containers to shield against sunlight.
     - **UV-Resistant Materials:** Invest in containers made from UV-resistant materials to ensure long-term water quality.
   - Indoor Storage
     - **Opaque Containers**: Even indoors, use opaque containers or store water in a cool, dark place to prevent light-induced degradation.

3. **Pest Prevention**
   - **Insect and Rodent Control**
     - **Sealed Containers**: Use tightly sealed containers to prevent insects and rodents from contaminating the water.
     - **Elevated Storage:** Elevate containers to deter rodents and use traps or repellents in storage areas.
   - **Filtering**
     - **Mesh Screens:** Place mesh screens or filters over openings to prevent insects and debris from falling into your container.
     - **Chemical Treatments:** Consider using safe chemical treatments that deter pests without compromising water safety.
4. **Container Selection**
   - **Material Considerations**
     - **Food-Grade Containers**: Always use food-grade containers to avoid leaching harmful chemicals into the water.
     - **Plastic vs. Metal**: Plastic is lightweight and convenient but may degrade over time. Metal containers are durable but may impart a metallic taste.
   - **Rotation and Inspection**
     - **Regular Rotation:** Rotate stored water regularly to ensure freshness and prevent stagnation.
     - **Inspect for Leaks:** Regularly check containers for leaks or damage that could compromise water quality.

When dealing with water storage, make sure you always consider your environment. If you're dealing with extreme temperatures, choose containers that can handle the heat or provide insulation against the cold. Opt for materials that maintain the quality of your water, and conduct regular checks to ensure everything stays in tip-top shape. Vigilance is key. Regularly inspect and rotate your water storage to prevent any surprises. Prioritize containers made from food-grade materials to ensure water purity. Think of your water supply as an investment in resilience rather than just a backup plan. Consider long-term strategies. Think about preservatives for extended storage, and explore crafting emergency water pouches at home for added preparedness. Functionality matters – choose containers with reliable taps, especially in accident-prone areas.

Chapter 10: Beyond the Bottle: The Many Uses for Water

Beyond its foundational role in hydration, water has versatile uses. From making personal hygiene more accessible to facilitating industrial processes, having access to water is a must. There are numerous examples where you'll learn that carrying out these processes becomes virtually impossible without water. Although humans have relied on water for agriculture and household needs for centuries, most industries require water, including food processing, chemical production, smelting, paper making, commerce, and much more.

Water in everyday life.
https://pixabay.com/photos/water-jet-shower-to-water-water-8007873/

The chapter digs into the practical aspects of leveraging water for survival, shedding light on its multiple dimensions as a crucial resource in the face of diverse challenges. This guide illuminates the role of water in fostering hygiene, emphasizing its significance in preserving health and well-being, especially when conventional medical resources may be scarce.

Water in Everyday Life

Hygiene and Personal Care

As you explore the many uses of water, hygiene stands out as a basic aspect of daily routines. It is essential for bathing, washing hands, and keeping the body clean. With adequate personal hygiene, harmful bacteria and microorganisms get washed away, protecting the body from diseases and further spread. If you're not maintaining hygiene, your body becomes susceptible to respiratory infections like colds and flu, diseases of the alimentary system may emerge, and even increases the chances of developing viral infections.

Cooking and Food Preparation

Besides its role as a simple ingredient, water becomes an essential component in the preparation and rehydration of food. While in survival situations and some cultures, cooking meals without water is a norm, most regions count water as an essential ingredient, making food palatable, nourishing, and easily digestible. Water is necessary for these cooking methods, whether boiling, simmering, or steaming. Most consumables can't be cooked properly without water or may lose nutritional value.

Cleaning and Sanitation

From cleaning the household to washing kitchen utensils, water is a powerful and natural ingredient that can be used for all your cleaning and sanitation needs. It is a universal solvent that removes dirt, grease, and other impurities, making your indoors squeaky clean.

Agriculture and Irrigation

It's the lifeblood of crops in agricultural practices. Water creates the correct soil moisture to promote seed sprouting root formation and supply essential nutrients. Water is also soaked up by the roots of already developed crops, promoting their healthy development and better yield. Plants can't grow, thrive, and produce an adequate yield without water.

It is impossible for plants to grow without water.
Photo by Tony Pham on Unsplash https://unsplash.com/photos/woman-in-black-long-sleeve-shirt-and-black-pants-standing-on-green-grass-field-during-daytime-TV7m_tpmqhw

Power Generation

As you already know, water is necessary for hydroelectric power generation. Hydropower plants harness the energy of flowing water to drive turbines that, in turn, generate electricity. It's one of the most reliable renewable energy sources, providing humans with power for decades.

Industrial Processes

Many modern-age industrial processes rely heavily on the use of water. It has numerous industrial applications, and it's widely used as a solvent in chemical reactions, for further processing of raw materials, in cooling systems that keep heavy machinery from overheating, and much more.

Construction and Building

Modern-day construction also relies heavily on water, essential in construction activities like mixing concrete, dust suppression, and settling foundations. It plays a critical role in creating infrastructure and buildings that form the backbone of communities.

Firefighting

Water is a primary tool in firefighting efforts. Fire hydrants, hoses, and relevant water sources are crucial for extinguishing fires and preventing their spread. Although fire-extinguishing substances like dry powders and carbon dioxide extinguishers are also used, various fire-extinguishing substances are water-based and use water in some form.

Understanding the diverse roles of water underscores its significance as a precious resource essential for sustaining life and supporting various aspects of human activities. However, with massive urbanization and irresponsible use of clean water, regions around the globe have started to face challenges regarding water supply. There's a growing need to update the current water supply and treatment systems to ensure the availability and conservation of water.

Water for Protection

In survival scenarios, where access to conventional hygiene practices may be limited, understanding how water is vital in personal cleanliness and wound care becomes imperative. Water becomes a crucial resource for first aid and emergency response. It's used for cleaning wounds, rehydrating individuals, and ensuring proper sanitation to prevent the spread of diseases. Here are some scenarios where the presence of water can make a difference.

Wound Cleaning

In medical emergencies and when giving first-aid, wound cleaning is the first step to preventing infection and promoting healing. Clean and uncontaminated water can effectively clean the site of injury, removing debris, harmful microorganisms, and dirt and priming the area for wound dressing. In medical terminology, rinsing the wound with clean or saline water is called wound irrigation, which reduces the risk of further infection and creates a suitable environment for the body's natural healing process.

Eye Irrigation

Like wound irrigation, eye irrigation is done when chemicals or foreign particles come into contact with the eyes. Water can flush out irritants and contaminants, providing rapid relief and preventing further damage. Continuous and gentle flushing removes every bit of the foreign substance. It minimizes the risk of injury as these particles can cause eye irritation, and when the eye is rubbed by force, it can damage the sclera and cornea, the outer layers of the eye.

Burn Treatment

Water is a fundamental component in the initial treatment of burns. Cool running water is applied to the burned area to help dissipate heat, reduce pain, and minimize tissue damage. Prolonged water application aids in cooling the burn and provides immediate relief. This immediate response with water is instrumental in the early stages of burn management when you are still waiting for professional medical attention. However, avoid using cold water or solid ice as it can exacerbate the burn site further.

Heat Exhaustion Mitigation

During hot days, heat exhaustion becomes a challenge that can be avoided by keeping yourself hydrated. In heat exhaustion, the body loses fluids, electrolytes, and minerals through sweating. This loss can be replenished by drinking water. However, in severe cases of heat exhaustion, people can become drowsy and even unconscious. In situations like these, alert the relevant authorities for medical assistance. While you wait for assistance, use a dampening cloth with water and gently rub it on the forehead, arms, legs, and stomach. This technique promotes evaporative cooling, lowering the body temperature and alleviating symptoms of heat-related illnesses.

Hypothermia Management

Water is integral in both preventing and managing hypothermia. Avoiding wet clothing and keeping the body dry in cold environments are essential to prevent heat loss. In cases where hypothermia has already set in, gradual warming with warm water to raise body temperature safely is the first step. Controlled water use is crucial to balancing the need for warmth without causing thermal shock to the body.

Decontamination from Chemical Exposure

Water is a primary component for decontamination in scenarios involving exposure to harmful chemicals. Rinsing the affected areas with copious amounts of water dilutes and removes the chemical substances, reducing the risk of further harm. Quick and thorough decontamination is critical to minimize the impact of chemical exposure and prevent the absorption of toxic substances into the body.

Oral Hydration in Shock

For individuals experiencing shock, maintaining proper hydration is crucial. In medical terms, shock can be triggered by severe trauma, allergic reaction, blood loss from an injury, or a heatstroke. When a person goes into a state of shock, the blood pressure drops drastically. Oral rehydration in this scenario stabilizes blood pressure and supports overall circulatory function, reducing the risk of organ damage. A sudden drop in blood pressure can also make the person unconscious. Avoid giving water if you notice the patient is feeling dizzy.

Splint Activation

In survival situations when you don't have immediate access to healthcare services, using emergency splints for a broken bone becomes crucial. These splints require water for activation and proper application. Emergency splinting is a temporary measure, but it's highly effective in keeping the injured bone stable until professional medical assistance is available. Water's role in activating bandages ensures proper adhesion and fitting, contributing to the stability of injured limbs.

Insect Bite Relief

It's a common first-aid practice for avid adventurers and explorers to use water after insect bites or stings. Running cold water on the area alleviates pain and reduces swelling. The cooling effect of water gives immediate relief and acts as a natural remedy in outdoor or wilderness first-aid scenarios.

Knowing the varied applications of water in first-aid scenarios emphasizes its critical role in providing immediate care, mitigating the impact of injuries, and contributing to the overall well-being of individuals in distress. Water is a versatile and essential resource in the first-aid toolkit, whether used for wound cleaning, temperature regulation, or chemical decontamination.

The Critical Role of Water

Besides assisting in first aid scenarios, there are various survival situations where water use becomes critical. Here are some of the diverse roles water is assigned, highlighting its necessity and the positive outcomes it delivers, even in the face of adversity.

Firefighting in Survival Situations

In survival scenarios, having the knowledge and ability to control and extinguish fires is crucial for safety and resource preservation. While using fire extinguishers is easier, it's best used for small fires. In case of a rapidly spreading fire, utilizing every method to minimize the spread of fire becomes necessary. Water, whether sourced from nearby rivers, lakes, or streams, can be carried in emergency supplies, becoming a primary tool for firefighting. It's used to douse flames, prevent the spread of fires, and protect essential resources like shelter and food supplies. In wilderness survival, where escape routes may be limited, effective firefighting can ensure your safety and the protection of your resources.

Signaling for Rescue

Water's reflective properties can be strategically utilized for signaling purposes in survival situations where you are stranded on or near a water body. Creating disturbances on the water's surface, such as waving or creating ripples, enhances visibility for potential rescuers. Mirrors or shiny surfaces can also be oriented to catch and redirect sunlight, creating visible signals. This method significantly increases the effectiveness of distress signals, making individuals more conspicuous to search and rescue teams or passing aircraft.

Water-Based Navigation Tools

Bodies of water serve as natural navigation guides in survival situations. Understanding water currents, watching the flow of rivers, and recognizing coastal features become valuable navigational tools. These water navigation techniques have been widely used for decades, assisting navigators to reach their destination without using modern geo-location gadgets like GPS. These water features offer insights into the lay of the land, aiding in decision-making for optimal navigation routes. This innate connection between water and navigation becomes a navigational advantage in diverse terrains.

Habitat Expansion

Although water's direct role in several situations is critical, groundwater and streams in an area promote habitat diversity. Water bodies attract various forms of wildlife. In survival scenarios, the presence of a water body points to the presence of nearby food sources. When water levels in these lakes and streams drop, it affects the groundwater tables, making them drop, limiting the water supply, and putting the habitat and nearby human population at risk. While water usage is a necessity for humans, making persistent conservation efforts ensures the streams, lakes, and rivers stay flowing.

Water's significance in survival tasks goes beyond its conventional roles. It becomes a dynamic and adaptable resource, offering solutions for firefighting, signaling, navigation, and resource procurement. Understanding the diverse applications of water in survival scenarios can empower you to use its potential creatively, contributing to your overall resilience and success in navigating and overcoming challenges in the wild.

Water in Crafting

Clay Molding and Construction

Pottery and using clay for construction are ancient techniques still being used today. Knowing how to mold clay and construct a shelter using essential Earth elements like water and clay can become a lifesaver in survival or emergencies. Water makes the clay malleable, so it can be shaped intricately. This simple yet effective technique can create containers, tools, and shelter.

Plant Processing and Fiber Extraction

Besides clay molding, water is crucial in processing plants for crafting materials. In survival crafts, plants are often used for weaving baskets, creating cordage, or constructing shelters. Water softens plant fibers, making them more pliable and easier to manipulate. Processed plant material is often used with clay to reinforce the construction and make it more resistant to weather changes. Furthermore, this soaking process enhances the plant fiber's flexibility, making it easier to weave into functional items like baskets and fiber-based containers. In survival situations, weaved baskets or containers can carry water, store food, or organize belongings.

Natural Dyeing and Coloring

Traditional crafts involve the art of natural dyeing, where water serves as the medium for extracting colors from plant materials. These decades-old dyeing and coloring methods can camouflage clothing, create signaling flags, or cater to specific functions in survival situations. The use of natural dyes, facilitated by water, adds a layer of resourcefulness to crafting.

Tanning and Leatherworking

Water makes the processing of raw hides into leather much easier. In the wild, this craft has been used by remote communities for ages, transforming raw hides into durable leather. In survival scenarios, this skill allows individuals to process animal hides for clothing, footwear, or improvised containers. Water softens hides, making them more pliable for crafting, and is integral to the tanning process, preserving the hides and transforming them into usable leather.

Cordage and Rope Making

Crafting cordage or rope from plant fibers is another valuable skill in survival scenarios. Water is used to soften and twist fibers, making creating strong and durable ropes easier. These ropes serve various purposes, from constructing shelters to crafting tools. Mastery of cordage-making enhances resourcefulness, allowing individuals to improvise essential items for different tasks.

Traditional Cooking and Food Preparation

Water is indispensable in traditional cooking as a critical component for boiling, steaming, and stewing. In survival situations, knowing how to purify water for consumption and use it for cooking safely is essential. Water ensures that safe and nourishing meals can be prepared and facilitates the extraction of nutrients from edible plants, making them more palatable and digestible.

Fire Craft and Fire-Hardening Tools

Water is strategically involved in firecraft and tool hardening. Traditional fire-starting methods often utilize materials that benefit from water treatment, making them more conducive to friction-based ignition. Likewise, tools crafted from wood or bone can be hardened through controlled exposure to water and fire, enhancing their durability and functionality in survival scenarios.

Natural Medicine and Herbal Infusions

Water is fundamental in traditional medicine and herbalism, serving as a medium for extracting beneficial compounds from plants. In survival situations, knowledge of medicinal plants and the ability to create herbal infusions or poultices using water becomes vital. Water helps extract therapeutic properties, allowing individuals to address minor ailments and injuries using natural remedies.

Environmental Awareness and Weather Prediction

Water's behavior in the environment holds valuable information for survival. To predict weather changes, traditional skills involve reading signs in water bodies, such as ripples, currents, or aquatic life. Understanding these cues aids in planning and adapting to environmental conditions, maximizing the chances of successful survival. It exemplifies how water is not just a resource but a source of information crucial for adapting to the ever-changing conditions of the outdoors.

Water's role in traditional skills and survival crafts goes beyond its physical properties. It catalyzes creativity and adaptability, allowing anyone to craft essential items, extract colors, and predict environmental changes. By understanding and utilizing water in these diverse skills, individuals enhance their resilience and resourcefulness in navigating the challenges of survival scenarios.

No doubt water is the elixir of life, and without it, life's existence becomes impossible. It's a one-of-a-kind substance, a driving force to living things, and a crucial element that is required in every aspect of life, from crafting to survival. The Earth is covered by 71% water, of which only 3% is fresh water. Out of this 3%, only 1.2% of water is drinkable. Although this precious liquid fueling life on Earth is still accessible, population expansion and urbanization threaten clean drinking water reserves on the planet. Fortunately, most governments worldwide are now taking effective steps to control water wastage and work on water conservation by planting more trees and creating urban green spaces. Many regions have even started monitoring water usage to gather relevant information and to take effective action to prevent the loss of this precious life source.

Bonus Chapter: Checklists

Now that you have worked through the book, this checklist will help guide your decisions on your water survival. The checklist covers all aspects of water use, including storage, purification, the equipment you need, and what your water will be used for. Use these guidelines as a quick reference to ensure you are on the right path. Sometimes, a lot of dense information can be overwhelming. This checklist can help you sift through all the data to ensure you have not missed any essential considerations.

Tend to your water needs by keeping a checklist.
https://pixabay.com/photos/checklist-check-list-pen-3556832/

Whether you are setting up a homestead, traveling, prepping, or exploring the wilderness, an independent water source is needed. However, considering how dangerous it can be to consume and store water, you must double or triple-check that you are on point and have everything necessary for healthy, potable water. You do not want to find yourself in a position where you skim past an important detail that you already know. Even experts make mistakes, so having a comprehensive checklist to remind yourself of everything you must take into account is essential. Take your time and read through this checklist so that all your bases are covered and accounted for.

Hydro-Geography Checklist

- Have you considered the best way to collect water in your geographical location?
- What are the dangers affecting water in this area?
- How does the shape and structure of the land affect its water sources?
- Is the water in the area clean, or is it polluted?
- Is this region water scarce or abundant?
- Are there risks of flooding in the area?
- Are there seasonal droughts in the region?
- Does the water available in your area meet your daily requirements?
- What infrastructure or tools do you need to access the water?
- Is the water you can access on the surface or underground, and how will you gather it?

Rainwater and Dew Checklist

- How are you going to collect the rainwater?
- Which containers are best suited for your needs?
- What contaminants do you need to worry about in the area?
- Do you have a cool and dry place to store your water?
- Have you checked if your containers are leaking?
- Which water purification methods will you use to clean the rain or dew water?

- Is the surface you are using to collect rainwater clean?
- Will you boil your water or use other disinfection methods like bleach or chlorine tablets?
- Are you prepared for the dry season?
- Does your storage capabilities match your needs?
- Have you instituted cleaning protocols for yourself and the area where you store your water?
- Are your storage containers food-grade?
- Have you cleaned out your tanks?
- Do you rotate your water after six months?

Water Purification Checklist

- What kind of water filters are you using?
- Will you be making use of UV light technology to clean your water?
- Which kinds of chemical disinfection tablets meet your needs?
- Are your water purification methods portable so that you can carry them around?
- Have you considered how long you should boil your water by looking into the parasites and pathogens in your community or the region you are in?
- Have you checked for contaminants in your containers?
- Is your purification method suitable for what your water is being used for?
- Is your drinking water kept separated from water for other uses, and is it labeled?
- Do you have a water testing kit?

Snow and Ice Water Checklist

- Are you in an area where snow and ice are abundant?
- Is the ice made from freshwater or saltwater?
- Do you have enough fuel or energy to melt snow for your water consumption needs?
- What contaminants are in the snow, and how will you purify it?

Traveling with Water Checklist

- How much water do you need to carry, and do you have the capability to move that much?
- Are your containers durable?
- Do you have a way to boil the water?
- Have you got purification tablets?
- Where will you gather water when you travel, and is that source safe?
- Are your bottles or containers sealed properly?
- Have you got portable filtration devices?
- Do you have a UV purifier?
- Have you got an electric purifying device?
- Do you have a portable water testing kit?

Long-Term Water Storage Checklist

- Have your containers been used for anything other than water?
- Are your containers safe for water storage?
- Where will your water be stored, and what are you gathering water for?
- How long do you want to store water, and what material will the containers be made from that you will use?
- Do you have sufficient and appropriate space for the water?
- How many liters do you need to store?
- Does your water storage need to be mobile?
- How many people are making use of your water supply?
- Do you have the means to filter and disinfect your water?

Water Conservation Checklist

- Are your methods of water collection sustainable?
- Are the materials you are using environmentally friendly
- Have you been mindful of the chemicals you use and their effect on local ecology?
- Have you cleaned up after yourself?

- Have you responsibly used water without wasting it?
- Did you make considerations for the community around you when gathering water?
- Have you checked for leaks in your water storage and distribution systems?
- Did you consider your impact on the environment when you gather water?
- Are you using your fair share without overconsumption?

For all your water needs, work carefully through this checklist. Whether traveling with a small amount of water or storing hundreds of gallons over a long period, this checklist covers every aspect you should think about. The details of purification, storage, filtration, and distribution are too important to rush past. Take your time to work through this checklist and think deeply about what the prompts are pointing toward. Water can be the giver and taker of life, so it is important to respect this ancient substance. If you work well with water, it will meet many of your needs, but if you disrespect it, a disastrous outcome is inevitable.

Conclusion

Water is life. Humans, animals, birds, plants, and all other creatures need it to survive. If you plan to live off the grid, you must discover different water resources and learn how to collect, purify, and store the water for hydration, farming, bathing, etc.

The book starts by exploring the significance of water and its role in sustaining all life forms. You discovered how it is more vital than food and why it is always associated with life. You also learned the grave consequences of its deprivation.

You then understood the relationship between the physical characteristics of the land and the presence of water. You learned how to recognize mountain ranges, valleys, and other landforms. You uncovered how to locate water through its movement across landscapes.

You understood the natural process behind rain and dew formation. You then discovered the methodologies of collecting water through traditional and innovative techniques. Afterward, you learned to store the rain and dew water you collected. You learned about the best types of containers for long-term storage and their sizes and materials. You also saw the significance of keeping the water uncontaminated.

The collected water won't be safe for consumption until you purify it. You understood the risks of untreated water and learned different purifying methods so you and your family can consume clean water.

Most water resources will be frozen if you live in a cold environment. You discovered techniques for melting snow and ice. You also learned myths and misconceptions about snow, so you don't make any fatal mistakes.

If you live in an environment with limited water resources, you will need strategies to conserve water. You understood the challenges of a drought-prone region and learned effective techniques to minimize water usage.

People who are always on the move face more challenges than the ones who settle in one place. You learned about sourcing water when you are continuously mobile and your hydration needs due to increased physical exertion.

You then discovered techniques to keep your stored water supply uncontaminated and safe. You learned about the appropriate storage containers and their materials, designs, and capacities. You also discovered the best storage practices in various environments.

Water has many purposes besides hydration. You discovered the many uses of water in your everyday life. You also learned its significance in emergencies and survival tasks.

Life without water is impossible.

Here's another book by Dion Rosser that you might like

BUSHCRAFT FOR KIDS
Mastering the Art of Outdoor Survival and Thriving in the Wilderness

DION ROSSER

References

Cement, J. K. (2023, August 11). Rainwater Harvesting Methods, Techniques, and Tips. JK Cement. https://www.jkcement.com/blog/construction-planning/rain-water-harvesting-techniques/

Components of a Rainwater Harvesting System. (n.d.). Rainwaterharvesting.org. http://www.rainwaterharvesting.org/Urban/Components.htm

Harvesting Rainwater. (n.d.). Rainwater harvesting for drylands and beyond by Brad Lancaster. Rainwater Harvesting for Drylands and Beyond by Brad Lancaster. https://www.harvestingrainwater.com/

Housing News Desk. (2023, June 12). Rainwater Harvesting: Importance, Techniques, Pros, and Cons. Housing News. https://housing.com/news/different-rain-water-harvesting-methods/

Maxwell-Gaines, C. (2004, April 3). Rainwater Harvesting 101. Innovative Water Solutions LLC. https://www.watercache.com/education/rainwater-harvesting-101

Ogale, S. (2023). Rainwater Harvesting System. In Encyclopedia Britannica.

Rainwater Harvesting System: Steps, advantages & types. (n.d.). Ultratechcement.com. https://www.ultratechcement.com/for-homebuilders/home-building-explained-single/descriptive-articles/the-steps-to-an-efficient-rainwater-harvesting-system

Rainwater Harvesting. (2016, January 6). BYJUS; BYJU'S. https://byjus.com/biology/rainwater-harvesting/

Ruchi Singhal case study. (n.d.). Rainwater Harvesting. Cseindia.org. https://www.cseindia.org/rainwater-harvesting-1272

Sarkar, S. K., & Tigala, S. (2022, October 27). Harvest Rainwater for Water Security. BusinessLine. https://www.thehindubusinessline.com/opinion/harvest-rainwater-for-water-security/article66060897.ece

Vartan, S. (2020, December 4). A Beginner's Guide to Rainwater Harvesting. Treehugger. https://www.treehugger.com/beginners-guide-to-rainwater-harvesting-5089884

Water conservation : Rainwater Harvesting. (n.d.). Mygov.In. https://blog.mygov.in/water-conservation-rainwater-harvesting/

Revolutionizing urban spaces: 5 innovative rainwater harvesting techniques. (n.d.). Revolutionising Urban Spaces: 5 Innovative Rainwater Harvesting Techniques. https://smartwateronline.com/news/revolutionising-urban-spaces-5-innovative-rainwater-harvesting-techniques

(N.d.). Iwaponline.com. https://iwaponline.com/ws/article/20/8/3052/75992/Crafting-futures-together-scenarios-for-water

(N.d.). Masterclass.com. https://www.masterclass.com/articles/how-to-find-water

(N.d.-a). Amnh.org. https://www.amnh.org/explore/ology/water/what-is-water

(N.d.-a). Nyp.org. https://www.nyp.org/healthlibrary/definitions/untreated-water#:~:text=Untreated%20water%20is%20drinking%20water,and%20mild%20to%20severe%20illness.

(N.d.-b). Amnh.org. https://www.amnh.org/exhibitions/water-h2o--life/life-in-water/humans-and-water

(N.d.-b). Aspiringyouths.com. https://aspiringyouths.com/advantages-disadvantages/water-purifier/#google_vignette

(N.d.-c). Iwaponline.com. https://iwaponline.com/jwh/article/19/1/89/78374/Shungite-application-for-treatment-of-drinking

9 advantages of seawater desalination systems. (n.d.). Pure Aqua. Inc. https://pureaqua.com/blog/9-advantages-of-seawater-desalination-systems/

A solution to water scarcity. (n.d.). TREE AID. https://www.treeaid.org/blogs-updates/water/

Advantages & disadvantages of desalination. (n.d.). Brother Filtration. https://www.brotherfiltration.com/pros-and-cons-desalination/

Anderberg, J. (2016, April 20). How to find water in the wild. The Art of Manliness; Art of Manliness. https://www.artofmanliness.com/skills/outdoor-survival/how-to-find-water-in-the-wild/

Bramley, A. (2022, January 3). Is it safe to eat snow? Scientists say yes - with these caveats. NPR. https://www.npr.org/sections/thesalt/2016/01/23/463959512/so-you-want-to-eat-snow-is-it-safe-we-asked-scientists

Cal, R. (2019, March 16). 11 methods for off-grid water filtration and purification guide. Rustic Skills; Regina Cal. https://rusticskills.com/off-grid-water-systems/off-grid-water-filtration-purification/

Caldwell, J. (2019, July 11). Desalination Water Filtration Systems. What is Desalination? Water Equipment Technologies; Water Equipment Technologies - WET. https://wetpurewater.com/desalination-water-filtration-systems/

CDC. (2023, April 19). Creating and storing an emergency water supply. Centers for Disease Control and Prevention. https://www.cdc.gov/healthywater/emergency/creating-storing-emergency-water-supply.html

CDC. (2023a, April 13). Making water safe in an emergency. Centers for Disease Control and Prevention. https://www.cdc.gov/healthywater/emergency/making-water-safe.html

CDC. (2023a, April 13). Making water safe in an emergency. Centers for Disease Control and Prevention. https://www.cdc.gov/healthywater/emergency/making-water-safe.html

CDC. (2023b, April 13). Making water safe in an emergency. Centers for Disease Control and Prevention. https://www.cdc.gov/healthywater/emergency/making-water-safe.html

CDC. (2023b, April 19). Creating and storing an emergency water supply. Centers for Disease Control and Prevention. https://www.cdc.gov/healthywater/emergency/creating-storing-emergency-water-supply.html

Cho, R. (2011, March 7). The fog collectors: Harvesting water from thin air. State of the Planet; Columbia Climate School. https://news.climate.columbia.edu/2011/03/07/the-fog-collectors-harvesting-water-from-thin-air/

Clarke, J. (2022, December 10). Why can't you eat snow for hydration in a survival situation? advnture.com. https://www.advnture.com/features/dont-eat-snow

Complete guide to campervan water filtration. (2022, July 23). Engineers Who Van Life • DIY Van Building & Van Life. https://engineerswhovanlife.com/campervan-water-filtration/

Cunningham, R. (2023, May 2). How to collect dew water: Efficient techniques for natural resource utilization. Survival World. https://www.survivalworld.com/water/gathering-dew/

Debutify, & Tasneem, S. (2022, January 3). 7 amazing shungite water benefits (2023 update). Atmosure. https://atmosure.com/blogs/stories/shungite-water-benefits

Dew. (n.d.). Nationalgeographic.org. https://education.nationalgeographic.org/resource/dew/

Drinking-water. (n.d.). Who.int. https://www.who.int/news-room/fact-sheets/detail/drinking-water

DrinkPrime. (2023, April 10). Boiled Water vs. Filtered Water: Which is Better? Drinkprime.In; DrinkPrime. https://drinkprime.in/blog/boiled-water-vs-filtered-water/

Fink, L. (2022, August 3). Clay water pot - the best way to filter your water in 2022. Uai Central. https://uaicentral.com/blogs/news/clay-water-pot

Fitzgerald, S. (2019, March 7). 6 plastic-free ways to travel with safe drinking water. National Geographic. https://www.nationalgeographic.com/travel/article/how-to-drink-water-safety-on-vacation-sustainability

Grooved surface accelerates dew harvesting –. (2019, March 20). Physics World. https://physicsworld.com/a/grooved-surface-accelerates-dew-harvesting/

Haas, E. (2018, September 18). The Hiker's Guide to Staying Hydrated and Treating Dehydration. Backpacker. https://www.backpacker.com/survival/how-to-stay-hydrated-and-treat-dehydration/

Hari, A. (2023, October 20). Can eating snow dehydrate you? Truth or myth. Medium. https://medium.com/@marketing_14327/can-eating-snow-dehydrates-you-truth-or-myth-d00b4fdb3882

Harvesting water and harnessing cooperation: Qanat systems in the Middle East and Asia. (n.d.). Middle East Institute. https://www.mei.edu/publications/harvesting-water-and-harnessing-cooperation-qanat-systems-middle-east-and-asia

Hitchcock, J. (2023, March 3). How to boil water without electricity - 15 easy ways. Survival Stoic. https://survivalstoic.com/how-to-boil-water-without-electricity/

Home water storage for an emergency. (2017, December 8). Utah Department of Environmental Quality. https://deq.utah.gov/drinking-water/emergency-water-storage

How to desalinate water. (2011, May 16). wikiHow. https://www.wikihow.com/Desalinate-Water

How to find water in a survival scenario. (2021, October 18). Operatorsassociation.com. https://www.operatorsassociation.com/how-to-find-water-in-a-survival-scenario

How to find water in a survival situation. (n.d.). Tacticalgear.com. https://tacticalgear.com/experts/how-to-find-water-in-a-survival-situation

HOW TO harvest rainwater. (n.d.). Off-Grid Collective. https://www.offgridcollective.co.nz/pages/how-to-harvest-rain-water

Hung, E. (2018, October 14). Best water storage containers for emergencies [tested]. Pewpewtactical.com; Pew Pew Media, Inc. https://www.pewpewtactical.com/best-water-storage-containers/

Impulse, S. (n.d.). Solutions to water scarcity. Solarimpulse.Com; Solar Impulse Foundation. https://solarimpulse.com/water-scarcity-solutions

Individual & Family Health. (n.d.). Bacteria, viruses, and parasites in drinking water. State.Mn.Us. https://www.health.state.mn.us/communities/environment/water/contaminants/bacteria.html

Isobeld. (2016, February 25). Melt water: How to get water from snow and ice. TGO Magazine. https://www.thegreatoutdoorsmag.com/skills/melt-water-how-to-get-water-from-ice-and-snow/

KPS. (2017, April 21). 10 ways to find water to survive the wilderness. Know Prepare Survive. https://knowpreparesurvive.com/survival/10-ways-to-find-water/

Kresh, M. (2018, October 14). Can clay jugs filter water? Green Prophet. https://www.greenprophet.com/2018/10/how-clay-jugs-make-polluted-water-safe/

Kylene. (2018, August 8). How to store water for emergency preparedness. The Provident Prepper - Common Sense Guide to Emergency Preparedness, Self-Reliance and Provident Living. https://theprovidentprepper.org/how-to-store-water-for-emergency-preparedness/

Lewicky, A. (2008). How to Melt Snow for water. SierraDescents. https://www.sierradescents.com/2008/05/how-to-melt-snow-for-water.html

Libretexts. (2020, May 27). 13.3: Water Scarcity and Solutions. Biology LibreTexts; Libretexts. https://bio.libretexts.org/Bookshelves/Ecology/Environmental_Science_(Ha_and_Schleiger)/04%3A_Humans_and_the_Environment/4.02%3A_Water_Resources/4.2.03%3A_Water_Scarcity_and_Solutions

List the uses of water in our daily life. (2022, July 4). Byjus.com; BYJU'S. https://byjus.com/question-answer/list-the-uses-of-water-in-our-daily-life/

Managing Water Scarcity. (n.d.). World Wildlife Fund. https://www.worldwildlife.org/projects/managing-water-scarcity

McKay, K. (2021, August 22). How to store water for long-term emergencies. The Art of Manliness; Art of Manliness. https://www.artofmanliness.com/skills/outdoor-survival/hydration-for-the-apocalypse-how-to-store-water-for-long-term-emergencies/

Miller, K. (2021, January 1). Is it safe to eat snow? Doctors explain possible side effects - prevention. Prevention.com. https://www.prevention.com/health/a34618470/is-it-safe-to-eat-snow/

Millhone, C. (2023, October 24). Shungite: Is this 'healing' stone as good for you as people say it is? Health. https://www.health.com/shungite-7972956

Minimising algal growth in farm dams. (2021, March 9). Agriculture Victoria. https://agriculture.vic.gov.au/farm-management/water/managing-dams/minimising-algal-growth-in-farm-dams

Mollah, M. (2023, May 23). 7 Sustainable Solutions to Water Scarcity –. Sea Going Green.

Muskrat, J. (2017, October 16). Three ways to get safe drinking water from snow. Instructables. https://www.instructables.com/Three-Ways-to-Get-Safe-Drinking-Water-from-Snow/

Naude, J. (2022, November 1). How to Safely Store Water Long Term. Abeco Tanks. https://abecotanks.co.za/long-term-water-storage/

Offgrid Staff. (2017, July 26). The Myth of Water Rationing While Stranded in the Desert. RECOIL OFFGRID. https://www.offgridweb.com/preparation/the-myth-of-water-rationing-while-stranded-in-the-desert/

People, O., & Advisory Board. (2019, March 25). Understanding plastic recycling codes: Your guide to the RIC. Sustainable Brands. https://sustainablebrands.com/read/corporate-member-update/understanding-plastic-recycling-codes-your-guide-to-the-ric

Price, A. (n.d.). Obtaining water from snow and ice - dryad bushcraft. Dryadbushcraft.co.uk. https://www.dryadbushcraft.co.uk/bushcraft-how-to/obtaining-water-from-snow-and-ice

Rae. (2022, June 9). How to Make Fake Water for Crafts (+ Resin Terrarium Guide). Terrarium Tribe. https://terrariumtribe.com/fake-water-for-crafts/

Ramey, J. (2017, August 25). Best emergency water storage containers for your home. The Prepared. https://theprepared.com/homestead/reviews/best-two-week-emergency-water-storage-containers/

Ray, T. (2013, April 10). Water purification. American Hiking Society. https://americanhiking.org/resources/water-purification/

Reverse Osmosis. (2019, February 20). BYJUS; BYJU'S. https://byjus.com/chemistry/reverse-osmosis/

Reverse Osmosis. (2020, June 28). VEDANTU. https://www.vedantu.com/chemistry/reverse-osmosis

Rosinger, A. Y. (n.d.). Human evolution led to an extreme thirst for water. Scientific American.

Sahana. (2022, April 30). 10 advantages and disadvantages of chlorination of water to know. Tech Quintal. https://www.techquintal.com/advantages-and-disadvantages-of-chlorination-of-water/

Scavetta, A. (n.d.-a). Boiled water vs. Filtered water. Aquasana.com. https://www.aquasana.com/info/boiled-water-vs-filtered-water-pd.html

Scavetta, A. (n.d.-b). Water Filter vs. Water Purifier: What is the Difference? Aquasana.com. https://www.aquasana.com/info/water-filter-vs-water-purifier-pd.html

Shock chlorination. (2020, August 5). Well Water Program. https://wellwater.oregonstate.edu/well-water/bacteria/shock-chlorination

Shoop, M. (2010, September 10). How to filter water with clay pots. Sciencing; Leaf Group. https://sciencing.com/filter-water-clay-pots-6975650.html

Singh, P. K. (2023, September 27). Water purification and its advantages and disadvantages. Livpure. https://livpure.com/blogs/article/water-purification-and-its-advantages-and-disadvantages

Sissons, C. (2020, May 27). What percentage of the human body is water? Medicalnewstoday.com. https://www.medicalnewstoday.com/articles/what-percentage-of-the-human-body-is-water

SITNFlash. (2019, September 26). Biological Roles of Water: Why is water necessary for life? Science in the News. https://sitn.hms.harvard.edu/uncategorized/2019/biological-roles-of-water-why-is-water-necessary-for-life/

Smith, A. O. (2019, July 23). 10 effective ways to purify drinking water. A. O. Smith India. https://www.aosmithindia.com/easy-and-effective-ways-to-purify-water/

Solar water disinfection. (n.d.). Ctc-n.org. https://www.ctc-n.org/technologies/solar-water-disinfection

Solar Water Disinfection. (n.d.). Thewatertreatmentplants.com. http://www.thewatertreatmentplants.com/solar-water-disinfection.html

Solutions 12/01/2020, P. W. (2020, December 1). Is it safe to eat snow when thirsty? Pentair Water Solutions. https://www.pentair.com/en-us/water-softening-filtration/blog/snow-into-water.html#:~:text=Collect%20ice%20or%20snow%20and,to%20catch%20the%20falling%20water.

Store water safely - Hesperian health guides. (n.d.). Hesperian.org. https://en.hesperian.org/hhg/A_Community_Guide_to_Environmental_Health:Store_Water_Safely

Stricklin, T. (2023, October 5). 15 Dangerous Diseases Caused by contaminated drinking water. SpringWell Water Filtration Systems. https://www.springwellwater.com/15-dangerous-diseases-caused-by-contaminated-drinking-water/

Tallarico, G. (2018, July 20). Rainwater harvesting: 8 methods. World Permaculture Association. https://worldpermacultureassociation.com/rainwater-harvesting-8-methods/

The Best Ways to Stay Hydrated in the Wild. (n.d.). Survivor Filter. https://www.survivorfilter.com/blogs/home/the-best-ways-to-stay-hydrated-in-the-wild

TIMESOFINDIA.COM. (2023, August 25). Bringing back the Matka: Why clay pot water is the healthiest. Times Of India. https://timesofindia.indiatimes.com/life-style/food-news/bringing-back-the-matka-why-clay-pot-water-is-the-healthiest/articleshow/103057332.cms?from=mdr

Tom. (2023, February 2). Health risks of consuming melted snow water, what you should know. Weather Geeks.

Untreated water. (n.d.). Alberta.Ca. https://myhealth.alberta.ca/Health/Pages/conditions.aspx?hwid=stu3124&lang=en-ca

Us Epa, O. W. (2015). Emergency disinfection of drinking water. https://www.epa.gov/ground-water-and-drinking-water/emergency-disinfection-drinking-water

Us Epa, O. W. (2015). Emergency disinfection of drinking water. https://www.epa.gov/ground-water-and-drinking-water/emergency-disinfection-drinking-water

Us Epa, O. W. (2017). How we use water. https://www.epa.gov/watersense/how-we-use-water

UV Water Purification and How it Works. (n.d.). Espwaterproducts.com. https://www.espwaterproducts.com/understanding-uv-water-filtration-sterilization/

van Vuuren, A. (2023, June 28). The Advantages of Solar Powered Water Treatment Solutions. NuWater Water Treatment Solutions South Africa. https://nuwater.com/advantages-of-solar-powered-water-treatment-solutions/

Vuković, D. (2020, October 14). Food-safe plastics: Which plastic containers are safe for storing food and water? Primal Survivor. https://www.primalsurvivor.net/food-safe-plastics/

Water conservation initiatives in drought-prone regions. (n.d.). Energy5.

Water in the wild. (2020, May 11). The Survival University. https://thesurvivaluniversity.com/survival-tips/wilderness-survival-tips/all-things-water/water-in-the-wild/

Water purification and its advantages and disadvantages. (2018, April 23). Jplast.Co.Za; JPlast. http://jplast.co.za/2018/04/23/water-purification-and-its-advantages-and-disadvantages/

Water Scarcity. (n.d.). World Wildlife Fund. https://www.worldwildlife.org/threats/water-scarcity Lemire, M. (2023, November 3). 5 Tips to Prevent Dehydration While Hiking. Adventure Medical Kits. https://adventuremedicalkits.com/blogs/news/5-tips-to-prevent-dehydration-while-hiking

Water, C. (2022, December 13). What are the advantages and disadvantages of different water filtration methods? Clean Tech Water. https://www.cleantechwater.co.in/what-are-the-advantages-and-disadvantages-of-different-water-filtration-methods/

What are the benefits of a reverse Osmosis water filter? (n.d.). Espwaterproducts.com. https://www.espwaterproducts.com/reverse-osmosis-advantages-and-disadvantages/

What is Hydrogeology, and what do Hydrogeologists do? (2018, December 17). IAH - The International Association of Hydrogeologists. https://iah.org/education/general-public/what-is-hydrogeology

Why is water essential to life? (2020, January 9). Toppr Ask. https://www.toppr.com/ask/question/why-is-water-essential-for-life/

Williams, T. (2023, June 27). Your Survival Guide on how to find Water in the Wilderness. Desert Island Survival. https://www.desertislandsurvival.com/how-to-find-water/

Woodard, J. (2018, July 27). What is a reverse osmosis system and how does it work? Fresh Water Systems. https://www.freshwatersystems.com/blogs/blog/what-is-reverse-osmosis

Your guide to ion exchange water filters. (n.d.). Freedrinkingwater.com. https://www.freedrinkingwater.com/water-education/quality-water-filtration-method-ion-exchange.htm

Zagala, R. (2018, March 20). Best off-grid water filtration: How to filter your water without the power grid. The Berkey. https://theberkey.com/blogs/water-filter/best-off-grid-water-filtration-how-to-filter-your-water-without-the-power-grid

Zhou, W., Matsumoto, K., & Sawaki, M. (2023). Traditional domestic rainwater harvesting systems: classification, sustainability challenges, and future perspectives. Journal of Asian Architecture and Building Engineering, 22(2), 576–588. https://doi.org/10.1080/13467581.2022.2047979

Milton Keynes UK
Ingram Content Group UK Ltd.
UKHW020750260524
443077UK00005B/66